T0295238

BIO-ELECTROCHEMICAL SYSTEMS

This book addresses electro-fermentation for biofuel production and generation of high-value chemicals and biofuels using organic wastes. It covers the use of microbial biofilm and algae-based bioelectrochemical systems (BESs) for bioremediation and co-generation of valuable chemicals, including their practical applications. It explains BES design, integrated approaches to enhance process efficiency, and scaling-up technology for waste remediation, bioelectrogenesis, and resource recovery from wastewater.

Features:
- Provides information regarding bioelectrochemical systems, mediated value-added chemical synthesis, and waste remediation and resource recovery approaches.
- Covers the use of microbial biofilm and algae-based bioelectrochemical systems for bioremediation and co-generation of valuable chemicals.
- Explains waste-to-energy related concepts to treat industrial effluents along with bioenergy generation.
- Deals with various engineering approaches for chemicals production in an eco-friendly manner.
- Discusses emerging electro-fermentation technology.

This book is aimed at senior undergraduates and researchers in industrial biotechnology, environmental science, civil engineering, chemical engineering, bioenergy and biofuels, and wastewater treatment.

BIO-ELECTROCHEMICAL SYSTEMS

Waste Valorization and Waste Biorefinery

Edited by
Kuppam Chandrasekhar and
Satya Eswari Jujjavarapu

CRC Press
Taylor & Francis Group
Boca Raton London New York

CRC Press is an imprint of the
Taylor & Francis Group, an **informa** business

First edition published 2023
by CRC Press
6000 Broken Sound Parkway NW, Suite 300, Boca Raton, FL 33487-2742

and by CRC Press
2 Park Square, Milton Park, Abingdon, Oxon, OX14 4RN

© 2023 selection and editorial matter, Kuppam Chandrasekhar and Satya Eswari
Jujjavarapu; individual chapters, the contributors

CRC Press is an imprint of Taylor & Francis Group, LLC

ISBN: 978-1-032-12617-3 (hbk)
ISBN: 978-1-032-12619-7 (pbk)
ISBN: 978-1-003-22543-0 (ebk)

DOI: 10.1201/9781003225430

Typeset in Times
by MPS Limited, Dehradun

Contents

Preface

Electricity generation through the bioelectrochemical system (BES) has added importance in the current bioenergy situation because of its sustainable nature. A BES is a combination of biological and electrochemical processes to generate electricity, hydrogen, or any other useful chemicals and also helps in industrial wastewater treatment and supplies it back to the main stream. A BES works based on the new fermentation process called the electro-fermentation process, which is being developed to augment and enhance the potential of conventional fermentation methods. In this scenario, the proposed book is an attempt to address the veiled opportunities available through electro-fermentation in aiding the biofuel production. All of the chapters are written by established research professionals from different nations, shedding light on the recent trends and how this technology can be utilized to generate various high-value chemicals and energy using organic wastes. Bioelectricity is generated in microbial fuel cells (MFCs) under O_2-depleted surroundings, where microbes carry out set bioconversion reactions in order to change the carbon-rich organic wastes into electrons. Dedicated chapters on MFCs and state-of-the-art advancements, as well as their current limitations, are deliberated to ignite the minds of the readers. This book will also cover the use of microbial biofilm and algae-based bioelectrochemical systems for bioremediation and co-generation of valuable chemicals. A genuine review of the performance of this technology and its potential industrial applications will be presented. This book fills a pressing need to understand and transcend the knowledge gained in the area of electro-fermentation, which is expected to be the possible green alternative to the growing energy demands. The chapters in this book are organized with required tables and illustrations to reach a broad audience, mainly undergraduates, postgraduates, energy researchers/scientists, policy makers, and anyone else interested in the latest developments in electro-fermentation based on fuels and chemicals.

Dr. K. Chandrasekhar
Dr. J. Satya Eswari

Acknowledgments

Firstly, our greatest regards to the GOD, for bestowing upon us the courage, unfailing source of support, comfort, and strength to complete this book successfully.

We are greatly thankful to the director of National Institute of Technology Raipur and the head of department, Department of Biotechnology, National Institute of Technology Raipur for their continuous and unrelenting support.

We would also like to pay our regards to all the people, National Institute of Technology Raipur, our colleagues, mentors, friends, and family members for their emotional support.

Editors' Biographies

Dr. Kuppam Chandrasekhar has a Ph.D. in biological science from the Academy of Scientific and Innovative Research (AcSIR), Indian Institute of Chemical Technology (CSIR-IICT), Hyderabad, Telangana, India. He has worked as a senior research fellow at SIR-IICT. He is currently working as a research professor in the Department of Civil and Environmental Engineering, Yonsei University, South Korea. His main areas of research are bioenergy from waste biomass, microbial fuel cells, bihydrogen, and waste remediation technologies. He has published 54 papers in scientific journals, such as (1) *Bioresource Technology*, (2) *International Journal of Hydrogen Energy*, (3) *Renewable and Sustainable Energy Reviews*, (4) *Waste Management*, (5) *Journal of Biotechnology*, (6) *International Journal of Molecular Sciences*, (7) *Journal of Hazardous Materials*, (8) *Energies*, (9) *Chemosphere*, and (10) *Critical Reviews in Biotechnology*. He has authored more than 50 book chapters. His work has been cited 3,400 times with an h index of 29 and an i10 index of 42. He is a life member of the following scientific societies: (1) life member of The Biotech Research Society of India (BRSI), (2) senior membership in the Asia-Pacific Chemical, Biological & Environmental Engineering Society (APCBEES), (3) senior member in the Institute of Research Engineers and Doctors, and (4) frontiers member. He is serving as a review editor in *Bioenergy and Biofuels*. Then, he was a guest faculty, Department of Engineering, University of Naples "Parthenope," Naples, Italy. He has also presented his work at several national and international conferences.

Dr. Satya Eswari Jujjavarapu is currently working as an assistant professor in the Department of Biotechnology at the National Institute of Technology (NIT), Raipur, India. She did her M.Tech in biotechnology and biochemical engineering from Indian Institute Technology (IIT) Kharagpur and Ph.D. in chemical engineering from IIT, Hyderabad. During her research career, she worked as a woman scientist for a Department of Science and Technology (DST) project at the Indian Institute of Chemical Technology (IICT) Hyderabad. She has rigorously pursued her research in the areas of bioinformatics and bioprocess and product development. She gained pioneering expertise in the application of mathematical and engineering tools in biotechnological processes. Her fields of specialization include bioinformatics, biotechnology, and process modeling; evolutionary optimization; and artificial intelligence. She has more than 67 publications in SCI/Scopus-indexed journals and 58 proceedings in international and national conferences. Her research contributions have received wide global citations. She has also published four book chapters and four books (currently in press) with international publishers. She completed one DST woman scientist project (22 lakhs) and is currently handling two projects, a DST-early career research project (43 lakhs), and a CCOST (Chhattisgarh Council of Science and Technology) project (4 lakhs). She has a teaching experience of more than 7 years, along with a research experience of 3 years. She has published 4 books and 10 book chapters. She has conducted various short-term courses on bioprocess development, and soft computing and intelligent techniques. She has delivered invited lectures at various institutes. She visited three countries for international conferences. She has guided more than 55 B.Tech and post-graduate students and is currently supervising 6 Ph.D. students, a project assistant, and 7 B.Tech students. She is an active participant in both academic and research activities. She has reviewed various research articles in 20 international journals and is an active member of Indian Institute of Chemical Engineers, The Indian Science Congress Association, Indian Institute of Engineers, and Biotechnology Research Society and AMI. She has received an early career research (ECR) award by DST, IEI young engineer award, Venus Outstanding woman in engineering, and DK international best faculty award.

Contributors

Shadab Ahmed
Department of Biotechnology (merged
  with Institute of Bioinformatics and
  Biotechnology)
Savitribai Phule Pune University
Pune, India

Aarti Bisht
Department of Chemical Engineering
Konkuk University
Seoul, Republic of Korea

K. Chandrasekhar
School of Civil and Environmental
  Engineering
Yonsei University
Seoul, Republic of Korea

Hany Abd El-Raheem
Zewail City of Science and
  Technology
Giza, Egypt
and
Université Paris-Saclay
CNRS, Institut de Chimie Moléculaire
  et des Matériaux d'Orsay (ICMMO)
ECBB Université Paris-Saclay
Orsay, France

Saurav Gite
Department of Biotechnology (merged
  with Institute of Bioinformatics and
  Biotechnology)
Savitribai Phule Pune University
Pune, India

Rabeay Y.A. Hassan
Zewail City of Science and Technology
Giza, Egypt
and
Applied Organic Chemistry Department,
  National Research Centre (NRC),
  Dokki
Giza, Egypt

Pradnya Jadhav
Department of Biotechnology (merged
  with Institute of Bioinformatics and
  Biotechnology)
Savitribai Phule Pune University
Pune, India

Sanath Kondaveeti
Department of Chemical Engineering
Konkuk University
Seoul, Republic of Korea

Sakshi Kor
Department of Biotechnology (merged
  with Institute of Bioinformatics and
  Biotechnology)
Savitribai Phule Pune University
Pune, India

Hafsa Korri-Youssoufi
CNRS, Institut de Chimie Moléculaire
  et des Matériaux d'Orsay (ICMMO)
ECBB Université Paris-Saclay
Orsay, France

Jung-Kul Lee
Department of Chemical Engineering
Konkuk University
Seoul, Republic of Korea

Satya Sundar Mohanty
Department of Biotechnology
Karunya Institute of Technology and
    Sciences
Coimbatore, TN, India

Swati Sambita Mohanty
National Institute of Technology
Rourkela
Odisha, India

Raviteja Pagolu
Department of Chemical Engineering
Konkuk University
Seoul, Republic of Korea

C. Nagendranatha Reddy
Department of Biotechnology
Chaitanya Bharathi Institute of
    Technology (Autonomous)
Gandipet, Hyderabad,
    Telangana State, India

Dhanashri Satav
Department of Biotechnology (merged
    with Institute of Bioinformatics and
    Biotechnology)
Savitribai Phule Pune University
Pune, India

Devashish Tribhuvan
Department of Biotechnology (merged
    with Institute of Bioinformatics and
    Biotechnology)
Savitribai Phule Pune University
Pune, India

Vinay V
Department of Biotechnology (merged
    with Institute of Bioinformatics and
    Biotechnology)
Savitribai Phule Pune University
Pune, India

1 Conventional Anaerobic Digestion vs. Bioelectrochemical Treatment Technologies for Waste Treatment

Dhanashri Satav, Pradnya Jadhav, Sakshi Kor, and Shadab Ahmed

Department of Biotechnology (merged with Institute of Bioinformatics and Biotechnology), Savitribai Phule Pune University, Pune, India

CONTENTS

DOI: 10.1201/9781003225430-1

1

1.1 INTRODUCTION

What will be the source of energy/fuel in the near future once all the fossils/ petroleum are exhausted? The alternative to these is an electrical source of energy. With the focus now shifting to the use of renewable resources for energy or fuel production, the development of new advanced technologies to convert waste into useful energy or fuel becomes even more imminent. What if we combine the two problems and find a single solution? And the solution to this is generating energy in the form of electricity, biofuels, biomethane, biohydrogen, etc. from the waste material. The cost of being dependent on the raw material for generating energy is always high; therefore, a suitably treated waste can serve as a substitute to the conventional raw material to generate energy. Conventional anaerobic digestion (AD) was first demonstrated in the 17th century (the early 1630s) by a Belgian chemist, Jan Baptita Van Helmont. He showed that combustible gases can be obtained by de-composting organic matter. The first sewage plant was built in Bombay in the year 1859. In 1895, England designed a process to recover flammable gases by treating sewage. Two years later, the first biogas plant was set up in Bombay, India. Then, later in the 1930s, the grange waste was used to generate flammable gas to power the street lights of asylum in Bombay. The concept was soon used in the application and in the 1960s, Khadi and Village Industries Commission (KVIC) set up the biogas plant that can be used in rural areas as fuel for cooking and other domestic purposes (Muthudineshkumar & Anand, 2019). Alessandro Volta's experiment in 1776 showed that more amounts of combustible fuel can be produced, using more decaying organic matter. In 1895, England designed the sewage treatment facility and used the by-product generated to light the street lamps in Exeter. Until the early 1960s, China had set up a million biogas plants using the septic tank design as the basis and replaced the dome-shaped tank with a rectangular tank. India followed the same changes and participated in a Biogas Sector Partnership (BSP) along with Nepal and China (Muthudineshkumar & Anand, 2019). With increasing prices of oils and petroleum in the 1980s, the United Kingdom and Europe became interested in the biogas program as an alternative source of energy, which was renewable (Ismail et al., 2014; Wilkinson, 2011). With an increasing demand for energy and the importance of biogas, thus the setup of the first biogas plant in Bombay, India, researchers showed interest and made various modifications to the design of the biogas plant. Among all the plants, Grama Laxmi III was built by Joshbai Patel, which later became a guide for the KVIC floating dome model. The National Biogas and Manure Management program built up to 1,50,000 family-based biogas plants between the years 2009 to 2010 (Davis, 2005; Munasinghe & Khanal, 2010).

1.2 ANAEROBIC DIGESTION

Anaerobic digestion is a complex microbiological process in which many anaerobic and facultative bacteria work hand-in-hand/together to break down the complex organic matter into simple forms in anaerobic conditions (Munasinghe & Khanal, 2010, Parkin & Owen, 1986). Initially, the primary objective of anaerobic digestion of wastewater was for the utilization of organic matter, reduction in odor, and conversion of organic matter to methane and carbon dioxide. Thus, the biogas produced is an inexhaustible source of energy that can be utilized in various ways like producing heat, electricity, fuel boilers and furnaces, alternative to fuels for vehicles, and can also be used in households as natural gas pipelines. Today, biogas is cleaned and trace contaminants removed; thus, higher-quality gas is supplied as compressed natural gas (CNG) or liquefied natural gas (LNG). This can be more efficient for the internal combustion of engines and also used for domestic purposes. Anaerobic digestion is greatly used in many technologies, but it has a complex mechanism to understand since the biological factor "microorganisms" are involved, which are affected by slight changes in their environment like temperature, pH, moisture, etc. (Parkin & Owen, 1986; Nasir et al., 2012). The commonly used substrate for anaerobic digestion can be animal manure, food scraps, wastewater treatment solids, and municipal and industrial wastewater residues that are put into a digester to produce biogas (60% methane and 40% carbon dioxide) and digestate. The biogas produced can be purified to methane by removing CO_2, hydrogen sulfide, and other trace elements. The methane produced is used as an energy source/ biofuel (Raj et al., 2021a, 2021b). Another component formed is the digestate, which is rich in nutrients like nitrogen, phosphorus, and potassium. Hence, the digestate can be used as fertilizer and manure in fields. Thus, complete utilization of organic waste. The U.S. Energy Information Administration (EIA) estimated in 2019 that approximately 257 billion cubic feet of landfill gas was collected to produce about 10.5 billion kilowatt-hours (kWh) of energy; thus, supplying roughly around 0.3% of the total U.S. electricity consumption of 2019. EIA reports also suggest that in 2019, there are 65 waste treatment setups of anaerobic digestion for sewage and industrial wastewater, which can produce a total of 1 billion kWh of electricity. Also, dairy farms and livestock are rich sources of organic substrate for anaerobic digestion. The EIA estimates that for 2019, 25 large dairies and livestock operations in the United States generated around 224 million kWh of electricity from biogas (Table 1.1) (https://www.eia.gov/energyexplained/biomass/landfill-gas-and-biogas.php).

Anaerobic digestion (AD), as an energy-efficient method of disposing of organic waste and producing a high-value product at a lower cost, is widely used. Anaerobic digestion can be broadly divided into four major steps:

 i. Hydrolysis of complex organic compounds
 ii. Acidogenesis
 iii. Acetogenesis
 iv. Methanogenesis

1.2.1 CLASSIFICATION OF ANAEROBIC DIGESTION

The digesters of anaerobic digestion can be classified depending upon the type of feedstock, amount of moisture present in the substrate, and the temperature conditions. They can be categorized as follows:

TABLE 1.1
EIA Estimation of Energy Production in 2019

Source	Number of units involved	The energy produced (in kWh)
Landfill gas	336 landfills	10.5 billion
Sewage and industrial wastewater	65 types of the wastewater treatment facility	1.0 billion
Animal manure	25 large dairies and livestock operations	224 million

Source: U.S. Energy Information Administration.

I. **Batch and continuous system:** This type of anaerobic digestion depends on the mode of supply of raw material/substrate to the digester. If the feedstock once supplied and sealed, the biogas will be formed, but the rate of production of biogas will be more in the middle of the process as compared to that of the start and end of the process. Also, once all the feedstock is utilized, it has to be refilled to start the next cycle, thus known as the batch system. To increase the overall biogas production by maintaining a uniform biogas level throughout the process, the digester tank can be continuously supplied with the feedstock and constant agitation to produce large amounts of biogas continuously; thus, known as a continuous process for anaerobic digestion (Nizami & Murphy, 2010; Appels et al., 2011; Nasir et al., 2012; Rajendran et al., 2012; Matheri et al., 2016; Asato et al., 2019; Usack et al., 2012; Akil & Jayanthi, 2012).

II. **Wet and dry systems:** depending on the substrate solid concentration or moisture content of the substrate, they can be a wet system or dry system. The dry system consists of solid substrate concentration less than 20%–40%, which requires more energy input for transport and processing of thick slurry. Another system is a wet system consisting of less than 15% of solid substrate, requiring less energy input for transportation through the pumps and giving a higher amount of biogas production due to easy and quick exchange of nutrients of bacteria (Munasinghe & Khanal, 2010).

III. **Single-stage and multistage system:** When the anaerobic digestion is executed in a single sealed reactor, it is a single-stage system. But the disadvantage to this is that all the four main steps are being carried out in the same digester, thus due to acidogenesis the pH lowers and the methanogenic bacteria are hampered; thus, overall methane production decreases. This single-stage system is followed by a dry batch system or continuous wet system. Thus, to overcome this, we built a two-step or multistep system in which the first three processes i.e. hydrolysis, acidogenesis and acetogenesis are carried out in the first tank while the second tank is for methanogenesis of product from tank one. This system is ideal for continuous and wet process systems (Nizami & Murphy, 2010).

1.2.2 LIMITATIONS OF ANAEROBIC DIGESTION

Although, with the advancement of technology with respect to anaerobic digestion, there are certain operational limitations. Some of them are the low removal rate of nitrogen and phosphorous among other nutrients (Rajagopal et al., 2013; Gómez et al., 2019), reduced efficiency of treatment of wastewater at low temperature (Hansen et al., 1998, Patil et al., 2011), lower stability at high organic loading rates (Parkin & Owen, 1986), lower COD removal rates post starvation period (Ferguson et al., 2016), and inhibition of microbial growth due to the presence of biocides in the substrate (Kim et al., 2015).

1.3 BIOELECTROCHEMICAL TECHNOLOGIES (BETs)

With the advancement of time and technology, there is an increase in the complexity of the matrix of industrial wastewater; thus, it necessitates the need for a new strategy to combat the above drawbacks and increase the efficiency of such a treatment system (Xue et al. 2013; De Vrieze et al. 2014). Today, we have a potential technology to overcome the limitations of AD, i.e. microbial electrochemical technologies (METs) or bioelectrochemical technologies (BETs). This technology is built on the ability of electroactive bacteria (EAB) to transfer electrons to electrically conductive materials (Gajaraj et al. 2017; Logan et al. 2006).

Electroactive bacteria are a natural catalyst for the transfer of electrons to the electrodes so that bioenergy like electricity, biohydrogen, and methane can be produced (Butti et al., 2016; Logan, 2010). The electrons accepted by the anode are transported to either the cathode by an external electric circuit to generate electricity using a device called microbial fuel cells (MFCs) (Gajaraj et al., 2017; Logan, 2010; Logan, 2008) or to a counter electrode, under potentiostat control, in a device called a microbial electrolysis cell (MEC) (Pandey et al., 2016; San-Martín et al., 2019; Zhang & Angelidaki, 2014; Chandrasekhar et al., 2015). The primary application of the bioelectrochemical system (BES) is only the generation of electricity in combination with wastewater treatment using microbial fuel cells (MFCs) (Borjas et al., 2015). In a BES, either the anodic or the cathodic reactions are microbially catalyzed (Clauwaert et al., 2008). If the BES is generating electricity, then it is termed as microbial fuel cell (MFC) and if the BES is consuming electricity to catalyze the electrochemical reaction, then it is termed a microbial electrolysis cell (MEC) (Rabaey et al., 2007). In the early bioelectrochemical systems (BESs), the electron transfer between the microorganisms and the electrode was done via means of adding mediators (Rozendal et al., 2006; Allen & Bennetto, 1993; Kadier et al., 2017; Blake et al., 1994; Bond & Lovley, 2003). Thus, the development of a BES to generate the product via bioelectrochemical pathways has greatly expanded levels of bioenergy research. MFC and MEC, the major two variants of BESs, are used to generate bioelectricity, biohydrogen, nutrients, and minerals from the energy stored in wastewater (Clauwaert et al., 2007).

1.4 COMPONENTS REQUIRED TO SET UP A TREATMENT PLANT

1.4.1 ANAEROBIC DIGESTER

The design of the digester is successful when we consider the factors that include environmental conditions, quality and quantity of substrate, access to construction material, nature of subsoil, and availability of skilled labor (Chandrasekhar et al., 2021b; Enamala et al., 2019; Venkata Mohan et al., 2013; Venkata Mohan et al., 2011). Environmental conditions: Anaerobic digestion is a technology based on the working of microbes. Thus, utmost care must be taken to maintain their optimal surrounding temperatures, ranging from 30°C to 40°C; hence, advised for cooler locations or incorporation of temperature control in the design. Quality and quantity of substrate: The dimensions of the digester will depend on the quality and quantity of substrate that will be used in the plant. The components required for setup:

1.4.1.1   Complete Mix—Continuous Stirred Tank Reactor (CSTR)

i.   **Continuous stirred tank reactors** (CSTR) are relatively simple to design and operate as compared to other configurations and are widely used in solid waste treatments.
ii. In comparison with other configurations, CSTR has more uniformity in system parameters, such as temperature, mixing, chemical, and substrate concentration
iii. This reactor is efficient in the hydrolysis and acidogenesis of high-solid raw feedstock.

1.4.1.2   Up-Flow Anaerobic Sludge Blanket (UASB)

The up-flow anaerobic sludge blanket (UASB) is required to prevent the methanogen washout. This is done by separating the sludge retention time (SRT) from the hydraulic retention time (HRT) by the use of dense granular sludge. This technology is well established for high-strength wastewater rather than large solid particles. Thus, it has more suitable conditions for efficient acetogenesis and methanogenesis (12,13).

1.4.1.3   Anaerobic Sequencing Batch Reactor (ASBR) Configurations

The ASBR can be an alternative to CSTR and UASB. The ASBR operates in four cyclic steps, namely: feed, reaction, settling, and discharge. It allows for the decomposition of the substrate to methane and carbon dioxide by biological anaerobic metabolism. During the process, microbes are provided with a large amount of substrates, thus resulting in a high biomass production rate; hence, faster flocculation and settling. However, faster metabolism means a high amount of biomass production, thus leading to microbial washout at the beginning of the cycle. While at the end of the cycle, low substrate concentration leads to proper sludge settling and lower biogas formation. Thus, cycles should be frequently allowed to complete the four steps (14,15).

1.4.1.4   Plug Flow

The plug flow digester vessel is an insulated cylindrical tank made up of reinforced material such as concrete, steel, or fiberglass with a gas-tight seal to capture the biogas. It is operated at a temperature range of 20°C to 50°C without any internal agitation. It is fed with thick manure consisting of 11–14% total solids (https://extension.psu.edu/plug-flow-digester-vessel).

1.4.1.5   Covered Lagoon

A covered lagoon digester is a huge anaerobic lagoon with a high retention period. This system comprises 0.5–2% total solids discharge manure. The lagoon is covered with a flexible or floating gas-tight cover. They are naturally maintained at optimal temperatures for microbial growth and organic substrate digestion. The retention time is dependent on the size of the lagoon and can range from 30 to 45 days (https://extension.psu.edu/covered-lagoon).

1.4.1.6   Fixed Film

A fixed-film digester is a column packed with media like bagasse on which the methanogens grow and liquid substrate passes through the media. These digesters are also known as attached growth digesters. The methanogens grow on the media forming the biofilm. The retention time of fixed-film digesters can be less than five days.

1.4.2   Bioelectrochemical Systems or Setups

A BES consists of two half-cells, anodic and cathodic, which produce electricity or any other chemically derived products. Electrochemically active bacteria (EAB) utilize the organic substrate in the anodic chamber to generate electrons and protons. The electrons generated by the EAB are delivered to the anode. In order to maintain the system at electro-neutrality, protons (H^+) generated in the anodic chamber during the catalytic conversion travel through the cation exchange membrane (CEM) to the cathodic chamber (Clauwaert et al., 2007; Jadhav et al., 2017; Jadhav et al., 2017; Du et al., 2007; Aelterman et al., 2006; Choi & Ahn, 2013, Rosenbaum et al., 2005; Chae et al., 2008).

1.4.2.1   Electrodes

Electrodes play a vital role in the transfer of electrons (harvest the electrons in the anode chamber and get reduced in the cathode chamber) which consequences in the remediation of complex pollutants in a BET system. The anode material used is the same in MFC and MEC. These materials include carbon cloth, carbon paper, graphite-felt, granules, or brushes. Suitable electrode material should have few properties such as biocompatibility, no fouling nature, efficient electron discharge, high porosity, and sustainability/stability over a long period of time (Tharali et al., 2016; Velvizhi, 2019).

1.4.2.2   Chambers

Designs of construction of MFC depend on factors like the number of chambers, the mode of operation, and presence of the membrane. Thus, they can be classified as two-chamber MFCs (Du et al., 2007), single-chamber MFCs, and stacked MFCs (Aelterman et al., 2006; Choi & Ahn, 2013).

1.4.2.3 Membrane

For efficient operation of MFC technology, the two chambers need to be at electro-neutral. This is achieved with the help of proton exchange membranes (PEMs), which help in the transfer of protons between the chambers while keeping them isolated (Rosenbaum et al., 2005; Chae et al., 2008). The ideal characteristics of a PEM include cost, increased proton conductivity, good segregational properties, increased mechanical strength, thermal and chemical stability, and electronic stability (Oh & Logan, 2006; Rahimnejad et al., 2015; Peighambardousta et al., 2010; Bajpai, 2017).

1.4.2.4 Mediators

The electrons released during the metabolic activities of EAB require a mediator to be delivered to the electrode (Fultz & Durst, 1982). Here, the mediator extracts the electrons from the bacterial metabolic reactions and delivers to the anode electrode (Sevda & Sreekrishnan, 2012; Park & Zeikus, 2003). Some of the commonly used mediators are 2-hydroxy-1,4-napthoquinone, thionine, methylene blue, methyl orange, methyl red, etc. (Chae et al., 2008). Thus, all the components are used according to the required design in various combinations.

1.5 WORKING OF CONVENTIONAL ANAEROBIC DIGESTION AND BIOELECTROCHEMICAL TREATMENT

1.5.1 CONVENTIONAL ANAEROBIC DIGESTION

The four sequential steps involved in the process of conventional anaerobic digestion are hydrolysis, acideogenesis, acetogenesis, and methanogenesis (Anukam et al., 2019), as depicted in Figure 1.1.

a. **Hydrolysis** – The biomass is made up of complex polymers. A reaction happens due to the action of enzymatic activity of hydrolytic microorganisms which breaks down the complex polymers (carbohydrates, proteins, lipids) into the soluble organic molecules (sugars, amino acids, fatty acids).

b. **Acidogenesis** – The soluble organic molecules resulting from hydrolysis are converted by acidogenic bacteria to a mixture of alcohols, carbonic acid, volatile fatty acids (such as acetic, butyric, and propionic acids).

c. **Acetogenesis** – Acetogenesis is the third step in which the acetogenic bacteria convert the volatile fatty acids, alcohols, and carbonic acid to carbon dioxide, hydrogen, and acetate.

d. **Methanogenesis** – The third step provides the substrate for the last step i.e. methanogenesis. In this process, the acetate is converted into methane and carbon dioxide.

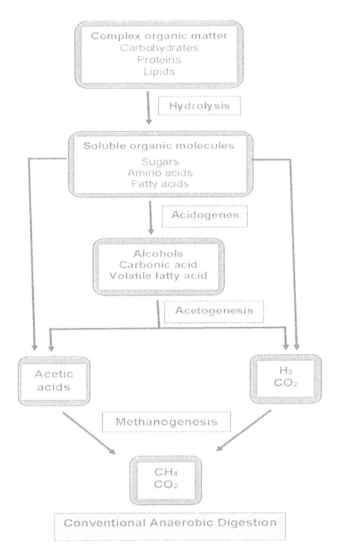

FIGURE 1.1 Scheme of conventional anaerobic digestion.

1.5.2 Working of Bioelectrochemical Treatment

Bioelectrochemical treatment generates electricity using microbial metabolism, where organic matter is broken down and the chemical energy is converted into electrical energy. As the organic waste or wastewater or sludge is acted upon by the microorganism, it generates an electrical half-cell, which maintains a reduction-oxidation system in the medium, subsequently resulting in an increase in the intensity and overall electrical capacity (Hernandez & Osma, 2020). The inorganic and organic compounds present in the wastewater are utilized by the microorganisms, where they perform a series of reactions called bioelectrochemical reactions and the process is known as "bioelectrochemical treatment" (BET) (Venkata Mohan et al., 2019).

1.5.2.1   Mechanism

The anode chamber of bioelectrochemical treatment is similar to conventional anaerobic treatment, both the reactions (oxidation and reduction) occur simultaneously (Venkata Mohan & Chandrasekhar, 2011a, 2011b; Nastro, 2014; Nastro et al., 2014). In the BET system, oxidation takes place at the anode (indirect anodic oxidation and direct anodic oxidation) and reduction takes place at the cathode (Gambino et al., 2021; Kumar et al., 2018; Chandrasekhar et al., 2014). On the surface of anode electrode of direct anodic oxidation, the pollutants get absorbed and due to anode electron transfer, the reactions get degraded (Chandrasekhar et al., 2021a, 2021c). After the introduction of artificial electrons, a potential difference is generated due to microbial metabolism, and the electrons are donated on the electrode to degrade the complex pollutants (Chandrasekhar & Venkata Mohan, 2012; Mohan et al., 2009; Mohanakrishna et al., 2010; Venkata Mohan et al., 2009). In indirect anodic oxidation, the electron exchange from organic pollutants takes place through some electroactive species produced during direct anodic oxidation (Mohan et al., 2009; Mohanakrishna et al., 2010; Venkata Mohan et al., 2009; Panizza & Cerisola, 2009). The concentration of primary oxidants is directly proportional to the concentration of secondary oxidants. The secondary oxidants produced also can be involved in pollutant degradation (Mohan et al., 2009; Israilides et al., 1997).

1.5.2.2   Working of Microbial Fuel Cells (MFCs)

In microbial fuel cells, at solid electrodes (electron acceptor), the organic matter gets oxidized by the microbes (Chandrasekhar et al., 2020). Oxidation takes place at the anode, where H^+ and electrons (e°) are generated and reduction takes place at the cathode. At the anode, the negative anodic potential is generated due to the presence of electrons, and a positive cathode potential is generated due to the movement of H^+ towards it by the proton exchange membrane.

$$\textbf{Anode:} \quad C_6H_{12}O_6 + 6H_2O \quad \rightarrow 6CO_2 + 24H^+ + 24e^-$$

$$\textbf{Cathode:} \quad 4e^- + 4H^+ + O_2 \quad \rightarrow 2H_2O$$

$$\textbf{Overall reaction:} \quad C_6H_{12}O_6 + 6H_2O + 6O_2 \rightarrow 12H_2O + 6CO_2$$

The performance of a MFC depends upon the electron transfer to the electrode surface from the bacterial cytoplasm. The electrons can directly get transferred on the electrode by physical contact without the involvement of redox species or mediators. Another way of electron transfer to the electrode is known as mediated electron transfer, which occurs through redox shuttles, through which the electron flow is mediated from bacteria towards the electrodes (Schröder, 2007).

1.5.2.3   Working of Microbial Electrolysis Cells (MECs)

In microbial electrolysis cells, there is involvement of a sealed cathode and external voltage and is similar to microbial fuel cells. This method was generated for hydrogen production from organic matter. The MEC is made up of a cathode chamber, anode chamber, and separator. At the anode, the organic matter is broken down to produce protons, electrons, and CO_2. To generate hydrogen, the electrons and protons move and reach the cathode through the external electric circuit and the electrolyte, respectively. The microbial biofilm (acts as an electrocatalyst) on the electrode supports the oxidation taking place at the anode (Chorbadzhiyska et al., 2011). The minimum overall cell voltage needed is shown as (Call & Logan, 2008) the potential needed to produce hydrogen at the cathode, $E_{CAT} = -0.41$ V. The anode potential of most MFCs reach around $E_{AN} = -0.30$ V.

Therefore, the minimum overall cell voltage needed is

$$\begin{aligned} E &= E_{CAT} - E_{AN} \\ &= (-0.41) - (-0.30) \\ &= -0.11 \text{ V} \end{aligned}$$

1.6   KINETICS AND THEIR PARAMETERS

1.6.1   KINETICS OF CONVENTIONAL ANAEROBIC DIGESTION

There are different kinetic models, such as first-order models (Zhen et al., 2015), logistic models and the modified Gompertz model (Zhao & Ruan, 2013) for the kinetic studies of conventional anaerobic digestion. These models can analyze the lag phase, hydrolysis rate, and biogas production rate and also biogas yield can be predicted. However, these kinetic parameters can be affected by the operational conditions (initial pH and substrate composition) and the process parameters (volatile fatty acids, total ammonia nitrogen, pH/alkalinity level) (Mao et al., 2017).

1.6.1.1   Kinetics of Bacterial Growth

The kinetics of microbial processes can be affected by the growth of microbes and the use of substrates. Under optimal conditions, the microbial growth takes place during the exponential phase while the acceleration and lag phases are negligible. At the exponential stage, the rate of bacterial growth is maximum and constant. The retardation phase starts when the nutrients exhaust, pH changes, and accumulation of toxic metabolites and the growth rate decline to zero. During the stationary period, the cell number remains constant, but cells consume energy due to biosynthetic processes or metabolism (Kythreotou et al.,

2014; Monod, 1949). If the medium conditions and growth rate do not change, the organisms die with a death rate (k_d). When both the growth and decline phases are at an exponential rate, the decline rate is smaller than the growth rate. When microbes die, they degrade into proteins and carbohydrates through the process of disintegration and will be consumed by viable microorganisms as substrates (Kythreotou et al., 2014). The Monod equation is usually used to epitomize bacterial growth kinetics (Giraldo-Gomez & Pavlostathis, 1991). This Monod equation relates the specific growth rate of bacteria to the concentration of the substrate as:

$$\mu = \frac{1}{X}\frac{dX}{dt} = \mu_{max}\frac{S}{S + K_s}$$

where, μ = Specific growth rate (time^{-1}), X = Concentration of active biomass (mass/volume), t = Time (time), S = Concentration of substrate (mass/volume), μ_{max} = Maximum specific growth rate (time^{-1}), and K_s = Saturation constant or half-velocity coefficient (mass/volume). At low substrate concentrations, the specific growth rate increases fast while the specific growth rate is slow at high substrate concentration. At low substrate concentrations, $S < K_s$ is first-order and at high substrate concentration $S > K_s$ is zero-order (Kythreotou et al., 2014).

1.6.1.2 Kinetics of Substrate Utilization

The growth rate of active biomass is correlated to the substrate utilization rate by the growth yield coefficient Y (g/g) (Giraldo-Gomez & Pavlostathis, 1991):

$$\frac{dS}{dt} = \frac{-(dX_a/dt)}{Y}$$

$$\frac{1}{X}\frac{dS}{dt} = \frac{\mu}{Y} = U$$

where S = substrate concentration (mass/volume) and U = specific substrate utilization rate (mass/mass/time). When μ is equal to μ_{max}, the ratio of μ_{max}/Y is maximum specific substrate utilization rate U_{max} (mass/mass/time). Therefore, the Monod equation for substrate utilization can be written as

$$r = \frac{-U_{max}S}{S + K_S}X_a$$

where r is the rate of substrate utilization (mass/volume/time).

1.6.1.3 Kinetics Studies for Batch Bioreactor

The growth requirements for microorganisms, substrate degradation, and gas production changes over the retention time in the batch processes (Giraldo-Gomez & Pavlostathis, 1991). Various studies showed the first-order model is used for substrate removal and biogas production for batch bioreactors

(Andualem et al., 2017; Antwi et al., 2017; Borja et al., 1993). The Gompertz and first-order models are most commonly used for the kinetic for batch bioreactor due to their simplicity (Córdoba et al., 2018).

1.6.1.4 Kinetics Studied for Continuous Bioreactor

The substrate endlessly transfers in and out of the reactor, ensuing constant substrate supply and biogas production in the continuous bioreactor. Hence, the growth necessities for microbes are continual (Kythreotou et al., 2014; Maleki et al., 2018). Modeling of various reactor configurations is done by integrating kinetic models and mass balances for substrate and biomass (Giraldo-Gomez & Pavlostathis, 1991; Batstone, 2006).

1.6.1.5 Effect of Temperature on the Kinetics of the Anaerobic Process

Temperature impacts biological processes by affecting the nutritional requirements, the nature of metabolism, reaction rates, and biomass composition (Esener, 1981). In various models, the influence of temperature on the anaerobic digestion process is measured by the Arrhenius equation. The equation shows that the variation in the natural log of the rate constant with temperature is proportional to the activation energy for the reaction (Rittmann & McCarty, 2001).

1.6.1.6 Effect of pH on the Kinetics of Anaerobic Process

The pH influences the development of bacteria and the kinetics of hydrolysis in anaerobic digestion. The optimal pH changes depending on the substrate and digesting method; therefore, the pH impact should be tuned correspondingly. The ammonia nitrogen buildup during substrate breakdown can impede biological function (Sánchez et al., 2000; Angelidaki & Ahring, 1993, Siegrist et al., 2002; Rosso et al., 1995).

1.6.2 Kinetics of Bioelectrochemical Treatment

BET analysis by Zhang et al. (Zhang et al., 2013) demonstrated that anodic denitrification-microbial fuel cell (AD-MFC) could be utilized for treating nitrate-containing wastewater and at the same time to generate electricity. A BET shows good treatment efficiency at pH 7 in terms of nitrates (33.5%/19.1%), COD (BET/AnT: 55%/51%), sulfates (58%/41%), and phosphates (33%/19%) in removal. Further treatment showed good color (100%/68%) and COD removal (BET/AnT, 95%/69%), which shows that the BET system can be a feasible platform to treat complex wastewaters with simultaneous energy recovery with an integrated approach (Li et al., 2014). Song reported in detail the kinetics and performance of a microbial fuel cell attached with synthetic landfill leachate (Song, 2017). They developed a microbial fuel cell that could convert the biochemical energy from the municipal solid waste (MSW) in addition to providing extended landfill longevity and minimized methane emission. One year later, Gadkari et al. (Gadkari et al., 2019) described a model for a single chamber microbial fuel cell (MFC), consisting of a biocatalyzed anode and an air-cathode.

It proved to be a great tool for a range of parameters and assists in typical process optimization. Also, in 2020, Zheng et al. (Zheng et al., 2020) comprehensively reviewed the developments in bioelectrochemical systems, the mode of electron transfer, and ultimately their applications.

1.7 ECONOMIC ANALYSIS

1.7.1 Economic Analysis of Conventional Anaerobic Digestion Systems

The most widely used process in conventional anaerobic digestion (CAD) is the treatment of sludge, which contains total solids (TS) of 3%–6% under mesophilic conditions (Appels et al., 2008). Thermophilic processes such as THSAD and TAD have a minimum environmental impact for common high-organic-content sludge. However, THSAD and THPAD (anaerobic digestion with thermal hydrolysis pre-treatment) show the best economic performance. THSAD has 44% less environmental impact and 118% higher net present value (NPV) for a project with a treatment capacity of 100t dry solids per day when compared with CAD. High-solid processes like THSAD and HSAD are preferred for low-organic-content sludge because these processes are much better than the others, mainly owing to less consumption of thermal energy. In comparison with CAD, THSAD can bring 40% minimum environmental burden and 31% more NPV in the case of this kind of feed sludge (Li et al., 2017). In China, the USA, and the United Kingdom, the high rate of activated sludge plus sludge anaerobic digestion process is recognized as better sustainable technology. In India, the flow anaerobic dislodge blanket plus activated sludge process is preferred in the present state scenario (Huang et al., 2019).

1.7.2 Economic Analysis of Bioelectrochemical Treatment

A microbial electrolysis cell (MEC) has been shown to contribute a significant amount of potential environmental impacts and cost of a bioethanol system. MEC current density is the central parameter for determining system size and electricity consumption. It has an impact on electricity usage due to electrode over-potentials, as well as system size due to the MEC surface area necessary to treat a constant wastewater flow. For the bioethanol system, it is observed that the economic viability mainly depends on MEC hardware costs and electricity price in the market, wastewater disposal, and CH_3OH. MEC hardware expenses can be reduced through systems engineering. The economic performance of MEC can be maintained by construction with low-cost materials as it is a large system. The low-cost materials are preferred over the system performance optimization by using expensive high-performance materials. System integration for an electrical serial connection of single MECs can be one of the better ways for significant cost reduction. If used under certain defined conditions, complementary application of methanol synthesis and microbial electrolysis, then it can be an environmentally and economically sound technology. In terms of environmental performance, the system can assist in mitigating some environmental effects. The construction material of the microbial electrolysis system, and as well as the system context are the main factors for determining the performance of the system (Streeck et al., 2018).

1.8 BY-PRODUCTS OF THE ABOVE TREATMENTS AND THEIR APPLICATIONS

Conventional anaerobic treatments produce a large number of by-products during the digestion process, and both the power and by-products produced are proven to be a great source of energy. Some of the important commercial and applied end products have been highlighted in Figure 1.2.

1.8.1 Biogas

Bechamp first reported methane formation from the decaying of organic matter due to the action of microorganisms in 1868 (https://extension.psu.edu/a-short-history-of-anaerobic-digestion). Volume per unit weight of volatile solids destroyed is expressed as gas production. The biogas is made up of 60%–70% methane and carbon dioxide and little amounts of H_2S, NH_3, N_2, H_2, etc. It is a

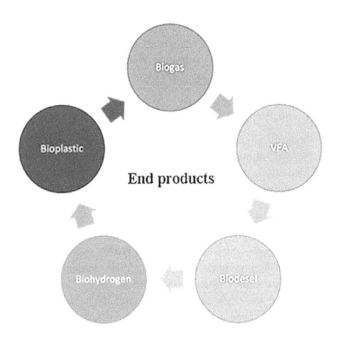

FIGURE 1.2 Potential value-added end products commercial importance that can be generated via anaerobic digestion (AD) and bioelectrochemical treatments (BETs).

renewable form of energy. Biogas production revolves around many aspects, such as BOD concentration, influent suspended solids concentration, and efficiency of the treatment process, temperature, pH, and VFA (Stein & Malone, 1980). Anaerobic digestion occurs via steps as mentioned earlier – hydrolysis, where the complex polymers of waste are hydrolyzed into simpler ones (monomers, sugars, amino acids). Then the second step is acidogenesis, the resulting monomers are converted to a combination of volatile fatty acid (butyric, propionic, and acetic acid) and trace amounts of hydrogen, and carbon dioxide by the action of acidogens. The third step is acetogenesis, where acetogenic bacteria act on the volatile fatty acids and convert them to acetate, hydrogen, and carbon dioxide. Methanogenesis is the fourth step; acetate is converted to methane, carbon dioxide, and hydrogen to methane gas (Murphy & Thamsiriroj, 2013; Chen & Neibling, 2014). Methanogenesis is a pH-dependent process and neutral pH is important for the production of methane. Hydrogen carbonate (bicarbonate alkalinity) is an important element for maintaining the pH of the solution for methanogenesis. Biogas is needed to be purified before use due to the presence of trace amounts of H_2S, water vapor removal, siloxane removal, and CO_2 removal. The purified biogas can be used for generating electricity, burning the boilers, and cooking and the purified biogas can be called biomethane ($CH_4 > 95\%$). Biogas is used to produce syngas by applying the Fischer-Tropsch process using various catalysts; the syngas can be converted to different liquid fluids or methanol (Stamatelatou et al., 2014).

1.8.2 BIOHYDROGEN

Hydrogen is a by-product produced during conventional anaerobic treatment. In conventional anaerobic digestion, the acidogenesis step can be used to produce hydrogen with some metabolites and VFA. In this step, the hydrogen can be produced by inhibition of hydrogen utilizing bacteria. Hydrogen can be effectively produced by regulating pH, substrate concentration, and temperature. Hydrogen is also known as an energy carrier. Useful in the production of hydrocarbon fuels (Stamatelatou et al., 2014). BET can be categorized into MFCs (waste remediation and generation of electricity) and MECs (generation of biomethane, biohydrogen). MECs are widely used for hydrogen production. At an anode in MECs, electrons, protons, and CO are generated due to oxidation by bacteria of substrates such as acetate. The electrons are carried to a solid anode by the electrochemical interaction of microbes. Due to the application of foreign voltage, the electrons drift to the cathode and combine with a proton to produce hydrogen gas. The protons drift from anode to cathode to make the charge neutral in the solution. In MECs, the voltage required for hydrogen production is approximately near to >0.2V (Logan et al., 2008).

$$\text{At anode: } CH_3COO + 4H_2O \rightarrow 2HCO_3^- + 8e - +9H^+$$

At cathode: $8H^+ + 8e^- \rightarrow 4H_2$ or $8H_2O + 8e \rightarrow 4H_2 + 8OH^-$

1.8.3 VOLATILE FATTY ACIDS (VFAs)

Volatile fatty acids such as acetic, propionic, butyric, isobutyric, valeric, and isovaleric is produced by acid-forming bacteria during the third step in the conventional anaerobic treatment process. There is a formula for estimating VFA production at a given point, (C2-C1)/(T2-T1) where C1 and C2 represent the initial and final concentration of VFA (g/l) and the T2 and T1 represents the duration of the study period (h) (Stein & Malone, 1980, Lata et al., 2002).

1.8.4 BIOPLASTICS

In the 1920s, the intracellular granules present in *Bacillus megaterium* were detected by the French microbiologist, Maurice Lemoigne, which are polyesters (poly[3-hydroxybutyrate], P[3HB}) and are classified among the polyhydroxyalkanoates (Lemoigne, 1926). The presence of PHA containing 3HB and 3-hydroxyvalerate in activated sludge was discovered by Wallen and Rohwedder in 1974 (Wallen & Rohwedder, 1974). PHA is synthesized by many archaea and prokaryotes through different renewable sources. PHA can be used in the pharmaceutical industry, medical, agro-industrial products, etc. After the breakdown of PHA, CO_2 and water are end products of the reaction, and they can be used by plants for their metabolism. PHAs are completely degrading polyesters (Akaraonye et al., 2010).

1.8.5 BIODIESEL

Biodiesel is a renewable form of biofuel. First-generation biofuels were generated using edible sources, which were sunflower, safflower, soybean, and rapeseed. Nonedible oil sources were applied for second-generation biodiesel production. Algae is used for biodiesel generation and solving rapid growth, land requirement, and CO_2 sequestration (Subhash & Mohan, 2015). Algae have an important property that is useful for biodiesel generation in combination with wastewater treatment that is the assimilation of carbon both heterotrophically and mixotrophically. As the wastewater is made up of both organic and inorganic matters, and the organic matters are composed of carbon, oxygen, and hydrogen, they can be used by the microalgae to produce biodiesel. The CO_2 is fixed during the day by the algae via photophosphorylation and carbohydrate is produced during the Calvin cycle (Liu & Benning, 2013).

1.9 CONCLUSIONS AND FUTURE PERSPECTIVES

Bioelectrochemical treatment systems are one of the emerging technologies for wastewater/organic waste/sludge remediation. The BET system has the advantage of treating high saline and recalcitrant wastewaters through the mechanism of direct and indirect oxidation methods. MECs are one of the new applications that use external voltage to generate hydrogen and methane. MFCs are applied on the low concentration substrate, whereas conventional anaerobic digestion can be applied on tons of wastewater remediation. In addition, we combine BETs with other processes to increase efficiency. The recent development in the micro in microbial electrochemical systems (MESs) has shown the tremendous potential of microorganisms to energetically interphase their environment using a diverse set of bioelectrochemical reactions. The development of electron transfer channels is a good example of how much potential an aptly chosen microorganism can deliver and how we can exploit its electroactivity for value addition and energy production (Zhang et al., 2019a; Zhang et al., 2019b).

REFERENCES

Aelterman, P., Rabaey, K., Pham, H. T., Boon, N., et al. (2006). Continuous electricity generation at high voltages and currents using stacked microbial fuel cells. *Environmental Science & Technology*, 40(10), 3388–3394.

Akaraonye, E., Keshavarz, T., & Roy, I. (2010). Production of polyhydroxyalkanoates: The future green materials of choice. *Journal of Chemical Technology & Biotechnology*, 85(6), 732–743.

Akil, K., & Jayanthi, S. (2012). Anaerobic sequencing batch reactors and its influencing factors: An overview. *Journal of Environmental Science and Engineering*, 54(2), 317–322.

Allen, R. M., & Bennetto, H. P. (1993). Microbial fuel-cells. *Applied Biochemistry And Biotechnology*, 39(1), 27–40.

Andualem, M., Seyoum, L., & Karoli, N. N. (2017). Kinetic analysis of anaerobic sequencing batch reactor for the treatment of tannery wastewater. *African Journal of Environmental Science and Technology*, 11(6), 339–348.

Angelidaki I., & Ahring B. (1993). Thermophilic anaerobic digestion of livestock waste: The effect of ammonia. *Applied Microbiology and Biotechnology*, 38, 560–564.

Antwi, P., Li, J., Boadi, P. O., Meng, J., et al. (2017). Efficiency of an upflow anaerobic sludge blanket reactor treating potato starch processing wastewater and related process kinetics, functional microbial community and sludge morphology. *Bioresource Technology*, 239, 105–116.

Anukam, A., Mohammadi, A., Naqvi, M., & Granström, K. (2019). A review of the chemistry of anaerobic digestion: Methods of accelerating and optimizing process efficiency. *Processes (MDPI)*, 7, 504.

Appels, L., Baeyens, J., Degrève, J., & Dewil, R. (2008). Principles and potential of the anaerobic digestion of waste-activated sludge. *Progress in Energy and Combustion Science*, 34(6), 755–781.

Appels, L., Lauwers, J., Degrève, J., Helsen, L., et al. (2011). Anaerobic digestion in global bio-energy production: Potential and research challenges. *Renewable and Sustainable Energy Reviews*, 15(9), 4295–4301.

Asato, C. M., Gonzalez-Estrella, J., Skillings, D. S., Castaño, A. V., et al. (2019). Anaerobic digestion of synthetic food waste-cardboard mixtures in a semi-continuous two-stage system. *Sustainable Energy & Fuels*, 3(12), 3582–3593.

Bajpai, P. (2017). Basics of anaerobic digestion process. In: P. Bajpai, *Anaerobic Technology in Pulp and Paper Industry*. Springer Briefs in Applied Sciences and Technology, pp 7–12.

Batstone, D. J. (2006). Mathematical modelling of anaerobic reactors treating domestic wastewater: Rational criteria for model use. *Reviews in Environmental Science and Biotechnology*, *5*(1), 57–71.

Blake, R. C., Howard, G. T., & McGinness, S. (1994). Enhanced yields of iron-oxidizing bacteria by in situ electrochemical reduction of soluble iron in the growth medium. *Applied and Environmental Microbiology*, *60*(8), 2704–2710.

Bond, D. R., & Lovley, D. R. (2003). Electricity production by *Geobacter sulfurreducens* attached to electrodes. *Applied and Environmental Microbiology*, *69*(3), 1548–1555.

Borja, R., Martín, A., Luque, M., & Durán, M. M. (1993). Kinetic study of anaerobic digestion of wine distillery wastewater. *Process Biochemistry*, *28*(2), 83–90.

Borjas, Z., Ortiz, J. M., Aldaz, A., Feliu, J., et al. (2015). Strategies for reducing the start-up operation of microbial electrochemical treatments of urban wastewater. *Energies*, *8*(12), 14064–14077.

Butti, S. K., Velvizhi, G., Sulonen, M. L., Haavisto, J. M., et al. (2016). Microbial electrochemical technologies with the perspective of harnessing bioenergy: Manoeuvring towards upscaling. *Renewable and Sustainable Energy Reviews*, *53*, 462–476.

Call, D., & Logan, B. E. (2008). Hydrogen production in a single chamber microbial electrolysis cell lacking a membrane. *Environmental Science and Technology*, *42*(9), 3401–3406.

Chae, K. J., Choi, M., Ajayi, F. F., Park, W. et al. (2008). Mass transport through a proton exchange membrane (Nafion) in microbial fuel cells. *Energy & Fuels*, *22*(1), 169–176.

Chandrasekhar, K., Amulya, K., & Venkata Mohan, S. (2014). Solid phase bio-electrofermentation of food waste to harvest value-added products associated with waste remediation. *Waste Management*, *45*, 57–65.

Chandrasekhar, K., Kumar, A. N., Raj, T., Kumar, G., Kim, S.-H. (2021a). Bioelectrochemical system-mediated waste valorization. *Systems Microbiology and Biomanufacturing*, *2021*(1), 1–12. https://doi.org/10.1007/S43393-021-00039-7

Chandrasekhar, K., Kumar, G., Venkata Mohan, S., Pandey, A., Jeon, B.-H., Jang, M., & Kim, S. H. (2020). Microbial electro-remediation (MER) of hazardous waste in aid of sustainable energy generation and resource recovery. *Environmental Technology & Innovation*, *19*(2), 100997. https://doi.org/10.1016/j.eti.2020.100997

Chandrasekhar, K., Lee, Y. J., & Lee, D. W. (2015). Biohydrogen production: Strategies to improve process efficiency through microbial routes. *International Journal of Molecular Sciences*, *16*, 8266–8293.

Chandrasekhar, K., Mehrez, I., Kumar, G., & Kim, S.-H. (2021b). Relative evaluation of acid, alkali, and hydrothermal pretreatment influence on biochemical methane potential of date biomass. *Journal of Environmental Chemical Engineering*, *9*, 106031. https://doi.org/10.1016/J.JECE.2021.106031

Chandrasekhar, K., Naresh Kumar, A., Kumar, G., Kim, D. H., Song, Y. C., & Kim, S. H. (2021c). Electro-fermentation for biofuels and biochemicals production: Current status and future directions. *Bioresource Technology*, *323*, 124598.

Chandrasekhar, K., & Venkata Mohan, S. (2012). Bio-electrochemical remediation of real field petroleum sludge as an electron donor with simultaneous power generation facilitates biotransformation of PAH: Effect of substrate concentration. *Bioresource Technology*, *110*, 517–525.

Chen, L., & Neibling, H. (2014). *Anaerobic digestion basics*. University of Idaho Extension, p. 6.

Choi, J., & Ahn, Y. (2013). Continuous electricity generation in stacked air cathode microbial fuel cell treating domestic wastewater. *Journal of Environmental Management*, *130*, 146–152.

Chorbadzhiyska, E., Hubenova, Y., Hristov, G., & Mitov, M. (2011). Microbial electrolysis cells as innovative technology for hydrogen production. *Natural Science*, 6, pp. 422–427. (Microsoft Word - Volume_1.doc (iaea.org)

Clauwaert, P., Aelterman, P., De Schamphelaire, L., Carballa, M., et al. (2008). Minimizing losses in bioelectrochemical systems: The road to applications. *Applied Microbiology and Biotechnology*, *79*(6), 901–913.

Clauwaert, P., Van der Ha, D., Boon, N., Verbeken, K., et al. (2007). Open air biocathode enables effective electricity generation with microbial fuel cells. *Environmental Science & Technology*, *41*(21), 7564–7569.

Córdoba, V., Fernández, M., & Santalla, E. (2018). The effect of substrate/inoculum ratio on the kinetics of methane production in swine wastewater anaerobic digestion. *Environmental Science and Pollution Research*, *25*(22), 21308–21317.

Davis, B. H. (2005). Fischer–Tropsch synthesis: Overview of reactor development and future potentialities. *Topics in Catalysis*, *32*(3), 143–168.

De Vrieze, J., Gildemyn, S., Arends, J. B., et al. (2014). Biomass retention on electrodes rather than electrical current enhances stability in anaerobic digestion. *Water Research*, *54*, 211–221.

Du, Z., Li, H., & Gu, T. (2007). A state-of-the-art review on microbial fuel cells: A promising technology for wastewater treatment and bioenergy. *Biotechnology advances*, *25*(5), 464–482.

Enamala, M. K., Pasumarthy, D. S., Gandrapu, P. K., Chavali, M., Mudumbai, H., & Kuppam, C. (2019). Production of a variety of industrially significant products by biological sources through fermentation. In: P. K. Arora (Ed.), *Microbial technology for the welfare of society* (pp. 201–221). Springer Singapore, Singapore. https://doi.org/10.1007/978-981-13-8844-6_9

Esener, A. A., Roels, J. A., & Kossen, N. W. F. (1981). The influence of temperature on the maximum specific growth rate of *Klebsiella pneumoniae*. *Biotechnology and Bioengineering*, *23*(6), 1401–1405.

Ferguson, R. M., Coulon, F., & Villa, R. (2016). Organic loading rate: A promising microbial management tool in anaerobic digestion. *Water Research*, *100*, 348–356.

Fultz, M. L., & Durst, R. A. (1982). Mediator compounds for the electrochemical study of biological redox systems: A compilation. *Analytica Chimica Acta*, *140*(1): 1–18.

Gajaraj, S., Huang, Y., Zheng, P., & Hu, Z. (2017). Methane production improvement and associated methanogenic assemblages in bioelectrochemically assisted anaerobic digestion. *Biochemical Engineering Journal*, *117*, 105–112.

Gadkari, S. Shemfe, M., Sadhukhan, J. (2019). Microbial fuel cells: A fast converging dynamic model for assessing system performance based on bioanode kinetics. *International Journal of Hydrogen Energy*, *44*(29), 15377–15386.

Gambino, E., Chandrasekhar, K., & Nastro, R. A. (2021). SMFC as a tool for the removal of hydrocarbons and metals in the marine environment: A concise research update. *Environmental Science and Pollution Research*, *28*, 1–16. https://doi.org/10.1007/s11356-021-13593-3

Giraldo-Gomez, E., & Pavlostathis, S. G. (1991). Kinetics of anaerobic treatment: A critical review. *Critical Reviews in Environmental Control*, *21*(5–6), 411–490.

Gómez, D., Ramos-Suárez, J. L., Fernández, B., Muñoz, E., et al. (2019). Development of a modified plug-flow anaerobic digester for biogas production from animal manures. *Energies*, *12*(13), 2628.

Hansen, K. H., Angelidaki, I., & Ahring, B. K. (1998). Anaerobic digestion of swine manure: Inhibition by ammonia. *Water Research*, *32*(1), 5–12.

Hernandez, C. A., & Osma, J. F. (2020). Microbial electrochemical systems: Deriving future trends from historical perspectives and characterization strategies. *Frontiers in Environmental Sciences, 8*(44), 1–19.

Huang, B. C., Li, W. W., Wang, X., Lu, Y. et al. (2019). Customizing anaerobic digestion-coupled processes for energy-positive and sustainable treatment of municipal wastewater. *Renewable and Sustainable Energy Reviews, 110*, 132–142.

Ismail, Z., Kong, K. K., Othman, S. Z., Law, K. H., et al. (2014). Evaluating accidents in the offshore drilling of petroleum: Regional picture and reducing impact. *Measurement, 51*, 18–33.

Israilides, C. J., Vlyssides, A. G., Mourafeti, V. N., & Karvouni, G. (1997). Olive oil wastewater treatment with the use of an electrolysis system. *Bioresource Technology, 61*(2), 163–170.

Jadhav, D. A., Ray, S. G., & Ghangrekar, M. M. (2017). Third generation in bioelectrochemical system research–a systematic review on mechanisms for recovery of valuable by-products from wastewater. *Renewable and Sustainable Energy Reviews, 76*, 1022–1031.

Kadier, A., Chandrasekhar, K., & Kalil, M. S. (2017). Selection of the best barrier solutions for liquid displacement gas collecting metre to prevent gas solubility in microbial electrolysis cells. *International Journal of Renewable Energy Technology, 8*, 93. https://doi.org/10.1504/IJRET.2017.086807

Kim, T. G., Yi, T., Lee, J. H., & Cho, K. S. (2015). Long-term survival of methanogens of an anaerobic digestion sludge under starvation and temperature variation. *Journal of Environmental Biology, 36*(2), 371.

Kumar, P., Chandrasekhar, K., Kumari, A., Sathiyamoorthi, E., & Kim, B. S. (2018). Electro-fermentation in aid of bioenergy and biopolymers. *Energies, 11*(2), p. 343.

Kythreotou, N., Florides, G., & Tassou, S. A. (2014). A review of simple to scientific models for anaerobic digestion. *Renewable Energy, 71*, 701–714.

Lata, K., Rajeshwari, K. V., Pant, D. C., & Kishore, V. V. N. (2002). Volatile fatty acid production during anaerobic mesophilic digestion of tea and vegetable market wastes. *World Journal of Microbiology and Biotechnology, 18*(6), 589–592.

Lemoigne, M. (1926). Products of dehydration and of polymerization of β-hydroxybutyric acid. *Bulletin in Society of Chemical Biology, 8*, 770–782.

Li, H., Jin, C., Zhang, Z., O'Hara, I. et al. (2017). Environmental and economic life cycle assessment of energy recovery from sewage sludge through different anaerobic digestion pathways. *Energy, 126*, 649–657.

Li, W. Z. Qian, Y. Chein-Chi Chang, M. J. (2014). Anaerobic process. *Water Environment Research, 87*(10), 1075–1094.

Liu, B., & Benning, C. (2013). Lipid metabolism in microalgae distinguishes itself. *Current Opinion in Biotechnology, 24*(2), 300–309.

Logan, B. E. (2008). *Microbial fuel cells* (pp 85–144). John Wiley & Sons, New Jersey.

Logan, B. E. (2010). Scaling up microbial fuel cells and other bioelectrochemical systems. *Applied Microbiology and Biotechnology, 85*(6), 1665–1671.

Logan, B. E., Call, D., Cheng, S., Hamelers, H. V., et al. (2008). Microbial electrolysis cells for high yield hydrogen gas production from organic matter. *Environmental Science & Technology, 42*(23), 8630–8640.

Logan, B. E., Hamelers, B., Rozendal, R., Schröder, U., et al. (2006). Microbial fuel cells: Methodology and technology. *Environmental Science & Technology, 40*(17), 5181–5192.

Maleki, E., Bokhary, A., & Liao, B. Q. (2018). A review of anaerobic digestion biokinetics. *Reviews in Environmental Science and Biotechnology, 17*(4), 691–705.

Mao, C., Wang, X., Xi, J., Feng, Y., et al. (2017). Linkage of kinetic parameters with process parameters and operational conditions during anaerobic digestion. *Energy*, *135*, 352–360.

Matheri, A. N., Mbohwa, C., Belaid, M., Seodigeng, T., et al. (2016). Design model selection and dimensioning of anaerobic digester for the OFMSW. In Proceedings of the *World Congress on Engineering and Computer Science* (Vol. 2).

Mohan, S. V., Raghavulu, S. V., Peri, D., & Sarma, P. N. (2009). Integrated function of microbial fuel cell (MFC) as bioelectrochemical treatment system associated with bioelectricity generation under higher substrate load. *Biosensors and Bioelectronics*, *24*(7), 2021–2027.

Mohanakrishna, G., Venkata Mohan, S., & Sarma, P. N. (2010). Bio-electrochemical treatment of distillery wastewater in microbial fuel cell facilitating decolorization and desalination along with power generation. *Journal of Hazardous Materials*, *177*(1–3), 487–494.

Monod, J. (1949). The growth of bacterial cultures. *Annual Reviews in Microbiology*, *3*(XI), 371–394.

Munasinghe, P. C., & Khanal, S. K. (2010). Biomass-derived syngas fermentation into biofuels: Opportunities and challenges. *Bioresource Technology*, *101*(13), 5013–5022.

Murphy, J. D., & Thamsiriroj, T. (2013). Fundamental science and engineering of the anaerobic digestion process for biogas production. In *The biogas handbook* (pp. 104–130). Woodhead Publishing.

Muthudineshkumar, R., & Anand, R. (2019). Anaerobic digestion of various feedstocks for second-generation biofuel production. In *Advances in eco-Fuels for a sustainable environment* (pp. 157–185). Woodhead Publishing.

Nasir, I. M., Mohd Ghazi, T. I., & Omar, R. (2012). Anaerobic digestion technology in livestock manure treatment for biogas production: A review. *Engineering in Life Sciences*, *12*(3), 258–269.

Nastro, R. A. (2014). Microbial fuel cells in waste treatment: Recent advances. *International Journal of Performability Engineering*, *10*, 367. https://doi.org/10.23 940/IJPE.14.4.P367.MAG

Nastro, R. A., Suglia, A., Pasquale, V., Toscanesi, M., Trifuoggi, M., & Guida, M. (2014). Efficiency measures of polycyclic aromatic hydrocarbons bioremediation process through ecotoxicological tests. *International Journal of Performability Engineering*, *10*, 411–418.

Nizami, A. S., & Murphy, J. D. (2010). What type of digester configurations should be employed to produce biomethane from grass silage? *Renewable and Sustainable Energy Reviews*, *14*(6), 1558–1568.

Oh, S. E. & Logan, B. E. (2006). Proton exchange membrane and electrode surface areas as factors that affect power generation in microbial fuel cells. *Applied Microbiology and Biotechnology*, *70*(2):162–169.

Pandey, P., Shinde, V. N., Deopurkar, R. L., Kale, S. P., et al. (2016). Recent advances in the use of different substrates in microbial fuel cells toward wastewater treatment and simultaneous energy recovery. *Applied Energy*, *168*, 706–723.

Panizza, M., & Cerisola, G. (2009). Direct and mediated anodic oxidation of organic pollutants. *Chemical Reviews*, *109*(12), 6541–6569.

Park, D. H., & Zeikus, J. G. (2003). Improved fuel cell and electrode designs for producing electricity from microbial degradation. *Biotechnol & Bioengineering*, *81*(3), 348–355.

Parkin, G. F., & Owen, W. F. (1986). Fundamentals of anaerobic digestion of wastewater sludges. *Journal of Environmental Engineering*, *112*(5), 867–920.

Patil, S. A., Harnisch, F., Koch, C., Hübschmann, T., et al. (2011). Electroactive mixed culture derived biofilms in microbial bioelectrochemical systems: The role of pH on biofilm formation, performance and composition. *Bioresource Technology*, *102*(20), 9683–9690.

Peighambardousta, S. J., Rowshanzamirab, S., Amjadia, M. (2010). Review of the proton exchange membranes for fuel cell applications. *International Journal of Hydrogen Energy*, *35*(17), 9349–9384.

Rabaey, K., Rodríguez, J., Blackall, L. L., Keller, J., et al. (2007). Microbial ecology meets electrochemistry: Electricity-driven and driving communities. *The ISME Journal*, *1*(1), 9–18.

Rahimnejad, M., Adhami, A., Darvari, S., Zirepour, A., & Oh, S. E. (2015). Microbial fuel cell as new technology for bioelectricity generation: A review. *Alexandria Engineering Journal*. 54(3):745–756.

Rajagopal, R., Massé, D. I., & Singh, G. (2013). A critical review on inhibition of anaerobic digestion process by excess ammonia. *Bioresource Technology*, *143*, 632–641.

Rajendran, K., Aslanzadeh, S., & Taherzadeh, M. J. (2012). Household biogas digesters-A review. *Energies*, *5*(8), 2911–2942.

Raj, T., Chandrasekhar, K., Banu, R., Yoon, J.-J., Kumar, G., & Kim, S.-H. (2021a). Synthesis of γ-valerolactone (GVL) and their applications for lignocellulosic deconstruction for sustainable green biorefineries. *Fuel*, *303*, 121333. https://doi.org/1 0.1016/J.FUEL.2021.121333

Raj, T., Chandrasekhar, K., Kumar, A. N., & Kim, S.-H. (2021b). Recent biotechnological trends in lactic acid bacterial fermentation for food processing industries. *Systems Microbiology and Biomanufacturing*, *2021*, 1–27. https://doi.org/10.1007/S43393-021-00044-W

Rittmann, B. E., & McCarty, P. L. (2001). *Environmental biotechnology: Principles and applications*. McGraw-Hill Education.

Rosenbaum, M., Schröder, U., & Scholz, F. (2005). In situ electrooxidation of photobiological hydrogen in a photobioelectrochemical fuel cell based *on Rhodobacter sphaeroides*. *Environmental Science & Technology*, *39*(16), 6328–6333.

Rosso, L., Lobry, J. R., Bajard, S., & Flandrois, J. P. (1995). Convenient model to describe the combined effects of temperature and pH on microbial growth. *Applied and Environmental Microbiology*, *61*(2), 610–616.

Rozendal, R. A., Hamelers, H. V., Euverink, G. J., Metz, S. J., et al. (2006). Principle and perspectives of hydrogen production through biocatalyzed electrolysis. *International Journal of Hydrogen Energy*, *31*(12), 1632–1640.

Sánchez, E., Borja, R., Weiland, P., Travieso, L., et al. (2000). Effect of temperature and pH on the kinetics of methane production, organic nitrogen and phosphorus removal in the batch anaerobic digestion process of cattle manure. *Bioprocess Engineering*, *22*(3), 247–252.

San-Martín, M. I., Sotres, A., Alonso, R. M., Díaz-Marcos, J., et al. (2019). Assessing anodic microbial populations and membrane ageing in a pilot microbial electrolysis cell. *International Journal of Hydrogen Energy*, *44*(32), 17304–17315.

Schröder, U. (2007). Anodic electron transfer mechanisms in microbial fuel cells and their energy efficiency. *Physical Chemistry Chemical Physics*, *9*(21), 2619–2629.

Sevda, S., & Sreekrishnan, T. R.(2012). Effect of salt concentration and mediators in salt bridge microbial fuel cell for electricity generation from synthetic wastewater. *Journal of Environmental Science and Health, Part A Toxic/Hazardous Substances and Environmental Engineering*, *47*(6), 878–886.

Siegrist, H., Vogt, D., Garcia-Heras, J. L., & Gujer, W. (2002). Mathematical model for meso- and thermophilic anaerobic sewage sludge digestion. *Environmental Science and Technology*, *36*(5), 1113–1123.

Song, Z. (2017). Characterization of kinetics and performance in a microbial fuel cell supplied with synthetic landfill leachate. *Open Access Master's Thesis*, Michigan Technological University, USA.

Stamatelatou, K., Antonopoulou, G., & Michailides, P. (2014). Biomethane and biohydrogen production via anaerobic digestion/fermentation. In *Advances in biorefineries* (pp. 476–524). Woodhead Publishing.

Stein, R. M., & Malone, C. D. (1980). Anaerobic digestion of biological sludges. *Environmental Technology*, *1*(12), 571–588.

Streeck, J., Hank, C., Neuner, M., Gil-Carrera, L. et al. (2018). Bio-electrochemical conversion of industrial wastewater-COD combined with downstream methanol synthesis-an economic and life cycle assessment. *Green Chemistry*, *20*(12), 2742–2762.

Subhash, G. V., & Mohan, S. V. (2015). Sustainable biodiesel production through bioconversion of lignocellulosic wastewater by oleaginous fungi. *Biomass Conversion and Biorefinery*, *5*(2), 215–226.

Tharali, A. D., Sain, N., & Osborne, W. J. (2016). Microbial fuel cells in bioelectricity production. *Frontiers in life science*, *9*(4), 252–266.

Usack, J. G., Spirito, C. M., & Angenent, L. T. (2012). Continuously-stirred anaerobic digester to convert organic wastes into biogas: System setup and basic operation. *JoVE (Journal of Visualized Experiments)*, *65*, e3978.

Velvizhi, G. (2019). Overview of bioelectrochemical treatment systems for wastewater remediation. In: S. Venkata Mohan, S. Varjani, & A. Pandey, *Microbial electrochemical technology* (pp. 587–612). Elsevier.

Venkata Mohan, S., & Chandrasekhar, K. (2011a). Self-induced bio-potential and graphite electron accepting conditions enhances petroleum sludge degradation in bioelectrochemical system with simultaneous power generation. *Bioresource Technology*, *102*, 9532–9541.

Venkata Mohan, S., & Chandrasekhar, K. (2011b). Solid phase microbial fuel cell (SMFC) for harnessing bioelectricity from composite food waste fermentation: Influence of electrode assembly and buffering capacity. *Bioresource Technology*, 102, 7077–7085.

Venkata Mohan, S., Chandrasekhar, K., Chiranjeevi, P., & Babu, P. S. (2013). Chapter 10 – Biohydrogen production from wastewater A2 – Pandey, Ashok. In: J.-S. Chang, P. C. Hallenbecka, & C. Larroche (Eds.), *Biohydrogen* (pp. 223–257). Elsevier, Amsterdam. 10.1016/B978-0-444-59555-3.00010-6

Venkata Mohan, S., Prathima Devi, M., Venkateswar Reddy, M., Chandrasekhar, K., & Asha Juwarkar, Sarma P.N. (2011). Bioremediation of real field petroleum sludge by mixed consortia under anaerobic conditions: Influence of biostimulation and bioaugmentation. *Environmental Engineering and Management Journal*, *10*(11), 1609–1616.

Venkata Mohan, S., Sravan, J. S., Butti, S. K., Krishna, K. V., et al. (2019). Microbial electrochemical technology: Emerging and sustainable platform. In S. V. Mohan, S. Varjani A. Pandey (Eds.), *Microbial electrochemical technology* (pp. 3–18). Elsevier, Amsterdam.

Venkata Mohan, S., Srikanth, S., Veer Raghuvulu, S., Mohanakrishna, G., Kiran Kumar, A., & Sarma, P. N. (2009). Evaluation of the potential of various aquatic ecosystems in harnessing bioelectricity through benthic fuel cell: Effect of electrode assembly and water characteristics. *Bioresource Technology*, *100*(7), 2240–2246.

Wallen, L. L., & Rohwedder, W. K. (1974). Poly-β-hydroxyalkanoate from activated sludge. *Environmental Science & Technology*, 8(6), 576–579.

Wilkinson, K. G. (2011). A comparison of the drivers influencing adoption of on-farm anaerobic digestion in Germany and Australia. *Biomass and Bioenergy*, 35(5), 1613–1622.

Xue, J. L., Liu, G. M., Zhao, D. F., Li, J. C. Z. et al. (2013). Inhibition effects of pentachlorophenol (PCP) on anaerobic digestion system. *Desalination and Water Treatment*, 51(28–30), 5892–5897.

Zhang, J., Zheng P., Zhang, M., Chen, H. et al. (2013). Kinetics of substrate degradation and electricity generation in anodic denitrification microbial fuel cell (AD-MFC). *Bioresource Technology*, 149, 44–50.

Zhang, T., Ghosh, D., & Tremblay, P. (2019a). Synthetic biology strategies to improve electron transfer rate at the microbe–anode interface in microbial fuel cells. In Krishnaraj, R. N., & Sani, R. K. (Eds), *Bioelectrochemical interface engineering* (pp. 187–208). John Wiley & Sons, Inc., Hoboken, NJ.

Zhang, Y., & Angelidaki, I. (2014). Microbial electrolysis cells turning to be versatile technology: Recent advances and future challenges. *Water Research*, 56, 11–25.

Zhang, Y., Liu, M., Zhou, M., Yang, H. et al. (2019b). Microbial fuel cell hybrid systems for wastewater treatment and bioenergy production: Synergistic effects, mechanisms and challenges. *Renewable and Sustainable Energy Reviews*, 103, 13–29.

Zhao, M. X., & Ruan, W. Q. (2013). Biogas performance from co-digestion of Taihu algae and kitchen wastes. *Energy Conversion and Management*, 75, 21–24.

Zhen, G., Lu, X., Kobayashi, T., Li, Y. Y., et al. (2015). Mesophilic anaerobic codigestion of waste activated sludge and *Egeria densa*: Performance assessment and kinetic analysis. *Applied Energy*, 148, 78–86.

Zheng, T., Li, J., Ji, Y., Zhang, W. et al. (2020). Progress and prospects of bioelectrochemical systems: Electron transfer and its applications in the microbial metabolism. *Frontiers in Bioengineering and Biotechnology*, 8(10), 1–10.

2 A Perspective on the Sustainable Bioenergy Production Coupled with Wastewater Treatment

Swati Sambita Mohanty
National Institute of Technology, Rourkela, Odisha, India

Satya Sundar Mohanty
Department of Biotechnology, Karunya Institute of
Technology and Sciences, Coimbatore, TN, India

CONTENTS

DOI: 10.1201/9781003225430-2

2.1 INTRODUCTION

India is the world's fourth-largest petroleum user, behind the United States, Japan, and China, with an annual impact of 6%–8% on the country's growing economy. It has the potential to have a significant effect on various forms of petroleum products (petrol, natural gases, or kerosene), as well as other environmental threats and global warming issues. The use of fossil fuels will continue to increase as energy demand rises significantly (Chandrasekhar et al., 2014). Fossil fuels currently account for most world's energy consumption, accounting for 86% of the total energy supply, leading to environmental pollution due to the CO_2 emitted during their usage (Alatraktchi et al., 2014; H. S. Lee et al., 2008). Fossil fuels are still plentiful and relatively inexpensive, but this is expected to change at some point (Chandrasekhar et al., 2021a). More particularly, its use is unlikely to be viable in the long term, owing to increased greenhouse gas (GHG) emissions as a result of their use (Chandrasekhar et al., 2015), as well as the environmental impact of such emissions on global warming (Hill et al., 2006). Dependency on it has become more environmentally challenging, and expensive (Chandrasekhar et al., 2021b,c). However, there is considerable interest in finding alternative sustainable fuel sources, potentially carbon-neutral, due to the depletion of the existing fossil fuels (Demirbas, 2009; Hill et al., 2006; Rittmann, 2008).

The best option to replace the existing energy sources is biogas, biofuels such as bioethanol, and biohydrogen. Biogas, a form of renewable bioenergy, has a composition of methane (45–70%), carbon dioxide (30–40%), nitrogen (1–15%), and traces of hydrogen sulphide as its main components, the constitution of which varies based on the source of origin (Hill et al., 2006). The most commonly synthesized and studied biofuels are bioethanol and biodiesel. The alcohol produced from the fermentation of carbohydrate-rich crops such as corn, sugarcane, etc. is known as bioethanol. Even the non-food sources of cellulosic biomass, such as grasses and trees, are also used as the source of the production of bioethanol. Microalgae is able to amass 100 times more oil components compared to its contemporary crop products (IEA, 2013). Similarly, renewable electricity or heat energy synthesized from biomass are known as bioelectricity and biopower, respectively.

Since 2000, world bioenergy production is estimated to increase sevenfold and still meet 2.3% of the final demands for liquid fuel (IEA, 2013). However, further study is highly necessary for several aspects of this field, such as reduction in operational costs or improvement in product yields as well as profitability, etc. (Kabir et al., 2018). Thus, in the current study, we have tried to discuss various strategies to enhance bioenergy production in a cost-effective manner with higher efficiencies.

2.2 SOURCES OF DIFFERENT FORMS OF BIOENERGY

To date, production of bioenergy has been carried out using a variety of sources, including both edible and non-edible crops, lignocellulosic material, agricultural residues, and algae. Microalgal biomass has gained immense popularity as a source of bioenergy over the last few decades (Raj et al., 2021a, 2021b, Venkata Mohan et al., 2011, 2013).

2.2.1 AGRICULTURAL WASTE

During various agricultural operations, a significant amount of dry biomass waste residue is being generated. These residues have usually been left as such to reduce soil erosion and allow nutrients to be recycled back into the soil. However, studies using these waste residues have been carried out to produce various energy complexes. Plants of the *Leguminosae* family, including *Dalbergia sissoo* (Indian rosewood), *Vachellia nilotica* (widely known as babul), *Perkia biglobosa* (locust bean), *Peltophorum pterocarpum* (yellow flame tree), and *Delonix regia* (flame tree), etc., could be utilized as the carbohydrate source. The seeds in these plant pods can be used as a carbohydrate source and a substrate in the fermentation process. These plant pods produce large quantities of reducing sugars after being treated with enzymes. When pods of *V. nilotica* are treated with 4% amylase, they make more reducing sugars, while *P. pterocarpum* has the least amount of reducing sugar.

Bioenergy can be derived from corn, maize, wheat, sugarcane, sorghum, Miscanthus, and other monocot plants. Corn-to-ethanol conversion technology is well known and well understood. The fermentation process is used to convert corn to ethanol. A large-scale ethanol plant will produce about 1 L of ethanol from 2.69 kg of corn grains. It takes approximately 100 cm of water to rinse the soil thoroughly in a single growing season. During the growing season of corn, the average irrigation needed for the total land is around 8.1 cm/ha. Corn stover is the better choice based on the costs of corn as a raw material. Stover is made up of leaves, husks, cobs, and stalk fractions. Corn stover could become a commonly used bioenergy product because of its widespread physical availability. It may be used in several ways, including raw material in bio- or thermochemical conversion processes to generate liquid fuels as the primary energy source in biomass cofiring applications (Zych, 2008). Maize is one of the most widely grown crops on the planet, and it has the unique ability to contribute to the production of biofuels. Maize must be produced for two purposes: grain production and stem biomass production, with higher yields, if used for biofuel generation. Maize can quickly be grown as a second crop due to agronomic and genomic resources. Maize has been regarded as the best model crop for biomass production due to all of these qualities and the availability of resources (van der Weijde et al., 2013). Wheat has a lot of promise as a biofuel crop. The production of ethanol from wheat using fermentation results in a fuel used to power automobiles. Wheat is classified as a C_3 plant species, which means it performs

C_3 photosynthesis. These plants can accumulate dry carbon mass, resulting in sufficient biomass for energy conversion (McKendry, 2002). Sugarcane is one of the most effective crops for absorbing solar energy and transforming it into chemical energy. The potential of sugarcane to be used as a biomass feedstock is well documented. When sugarcane is supplied for production, large quantities of sugarcane bagasse are generated, burned in boilers to produce steam and electricity. Advancements in bioethanol processing technologies create large quantities of bagasse, which can be used for various purposes, including electricity generation, bioethanol synthesis, and the development of different bio-based products (Cushion et al., 2009; van der Weijde et al., 2013). Sugarcane bagasse is a ligno-cellulosic material. The primary components of lignocellulosic products are cellulose, hemicellulose, and lignin. Glucose, mannose, xylose, and arabinose are the primary constituents of hemicellulose, which is a glucose polymer. Sugarcane bagasse must be refined into fermentable sugars before being used as a bioethanol feedstock (Elbehri et al., 2012). Sorghum contains two kinds of grains, namely the sugar type and biomass type, making it a unique species. Sorghum genome sequence also opens up possibilities for it to be used as a first- and second-generation biofuel crop. Forage sorghums are the most significant for biofuel production. The same techniques used to make sorghum can also be used to make biofuels from sweet sorghum. Sweet sorghum has a range of advantages over sugarcane, including abiotic stress resistance, resource quality, and better genetics, as well as the fact that it is an annual crop. Combining genetics, agronomic techniques, and processing technology will improve sorghum potential as a bioenergy crop (van der Weijde et al., 2013). C_4 grasses, especially those in the genus *miscanthus*, were identified as potentially improved bioenergy crops. The biomass production capacity of *Miscanthus* varies depending on the environment. *Miscanthus* x *giganteus*, a triploid hybrid, is the only plant that has been commercially grown in recent years. It is critical to consider the genus genetic diversity when optimizing the crop for various environmental conditions and increasing yield. *Miscanthus sinensis* and *Miscanthus sacchariflorus* are two possible biomass crops in the *Miscanthus* genus, both of which have a wide range in Asian tropical and temperate areas. *Miscanthus lutarioriparius*, a subspecies of *Miscanthus sacchariflorus* found in one part of China, can be used as a biomass crop. Domestication of *Miscanthus* is needed in the coming decades to make it a viable biomass crop (Asaoka & Atsumi, 2007; Robson et al., 2013; van der Weijde et al., 2013).

The use of edible oils as a feedstock for biofuel generation has many potentials. In recent years, soybean oil, rapeseed oil, and palm oil have accounted for 75% of total edible oil output. Biodiesel has been used in mineral diesel since the early 20th century but in smaller quantities. The use of biodiesel has increased dramatically in recent years, particularly since 2005, with the European Union (primarily France and Germany) playing a significant role in biofuel development, accounting for roughly 80% of global biofuel production. When it comes to biofuel production utilizing edible oils as feedstock, several factors must be considered, including the oil source (whether it is derived from food or non-food crops) and the oil composition and ability to serve as a feedstock. Despite biodiesel's immense potential, there are restrictions on how edible oils can be used as feedstocks due to rising demand and high costs (Calle et al., 2009). There are many methods for reducing vegetable oil viscosity and making them suitable for fuel, including micro-emulsification, pyrolysis, transester-ification, etc. Transesterification is the commonly used biodiesel generation process because that produces a high yield in a shorter reaction period at low temperatures and high pressure (Liaquat et al., 2012; Shikha & Rita, 2012).

Biogenic by-products, leftover parts, and waste products have all been successfully used as carbon sources for renewable energy sources (Nastro, 2014; Kadier et al., 2017; Kumar et al., 2018; Nastro et al, 2014). Various biomass sources have been considered for biofuel production (based on their utilization allocation). The biomass application allocation is affected by the final synthesizing scale, biofuel cost, and CO_2 reduction approach, with practical implications for improving the efficient methods of applying forest-origin biomass (Xin et al., 2010). Hence, bioenergy production can be future-proofed by maintaining efficient and sustainable biomass raw material existence (Gan & Smith, 2012).

2.2.2 MICROALGAE BIOMASS

Microalgae can be prokaryotic or eukaryotic and can be unicellular or simple multicellular organisms. They can effectively convert solar energy due to their simple cellular structure. Microalgae are thought to be among the planet's oldest living organisms. They are available in a wide range of shapes and sizes, with over 3,00,000 species. Many species of microalgae, despite their diversity, have an oil content of around 80%. Microalgae have the potential to be used as feedstocks for biodiesel production (Figure 2.1) (Saifullah et al., 2014; Htet et al., 2013). Microalgae are single-celled plant biomasses with a wide range of properties that could be used to produce liquid transportation fuels. These species can thrive in both freshwater and saturated saline (or both) environments, ef-fectively using CO_2 from the atmosphere and contributing more to atmospheric carbon fixation (up to 40%). Depending on the genus or species natural potential to generate energy-rich products (oils) in its overall dry biomass, this algal biomass can be made very quickly (doubling cycles 6 to 24 h). *Botryococcus* sp. has gathered up to 50% of its dry cell mass in long-chain hydrocarbons oil (Kojima & Zhang, 1999; Parker et al., 2008).

Researchers have various choices for identifying microalgae species bio-conversion using genetic sequences due to millions of algal species variants (Enamala et al., 2019). Compared to biofuel metabolism from terrestrial plant species, algal biofuel's biosynthesis capability will reduce fertile land utili-zation. Municipal wastewater treatment involves removing phosphates and nitrates before discharging them as effluents, which involves the possible use of microalgal organisms in waste streams (Douskova et al., 2009; Hannon et al., 2010).

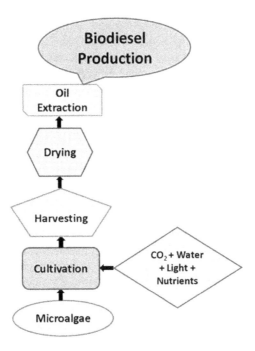

FIGURE 2.1 Different stages of microalgal biodiesel production.

2.3 POSSIBILITIES OF BIOENERGY PRODUCTION FROM SUSTAINABLE SUBSTRATES

2.3.1 BIOELECTRICITY GENERATION

Microbial fuel cells (MFCs) were being used for power generation (biosensors) in small-scale equipment (Chandrasekhar & Venkata Mohan, 2012; Gambino et al., 2021). The practical barriers to achieving minimum power with compact current flow in MFCs have been demonstrated. MFCs have been used as fuel in a variety of degradable organic substances under moderate operating conditions. Organisms can grow effectively, catabolizing substrates via bioelectricity generation (Rahimnejad et al., 2015). In both developing and developed countries, organic waste compounds recycling has become a significant challenge. Organic compost is made from waste organic compounds and has been used as a soil conditioner since ancient times. Organic compound conversion to electricity production was documented in several MFC designs and sizes and has recently gained more attention and emphasis. In this regard, compost-based MFCs were demonstrated to be capable of generating bioelectric power from organic substrates through oxidation and hydrolysis (Venkata Mohan & Chandrasekhar, 2011a, 2011b). Various organic waste products derived from leaves molds, rice grain interior covering (bran), oil cake from harvested mustard seeds, and carbon compound of chicken excreted substances have been provided to this MFC.

In the absence of oxygen, the electrical capacity (around 350 mV) is reported by affecting the different forms of membrane assembly, the mixing capability of fly ash materials, and the varied function of the electrode design process. Fly ash can be used to boost maximum voltage. A practical concept for optimizing power production was discovered using bamboo charcoal and carbon synthesized fiber as anode and cathode electrodes (Moqsud et al., 2013).

Food wastes (FWs) can be disposed of and converted into energy compounds using microbial fuel cells (MFCs). The loading rate of FWs affected MFC performance at a COD of 3.2×10^3 mgL^{-1}, and the highest power density of 18 mW/m^2 was found to be similar to 556 mW/m^2. It also has the maximum coulombic efficiency (CE) with a value of 27% of COD (4.9×10^3 mgL^{-1}). A combination of exoelectrogenic *Geobacter* and fermentative *Bacteroides* genera has been found to produce a more effective and durable MFC system with organic compound degradation and electric power synthesis activities (Jia et al., 2013). MFC systems for wastewater treatment with rapid power generation has demonstrated its design for bioenergy generation and wastewater remediation using currently advanced techniques such as bioelectrochemical systems (BESs). Food- and agriculture-based wastewater have been used by recognizing their elusive capability and discussing significant bottlenecks in improving process performance for sustainable energy recovery (ElMekawy et al., 2015). Organic fraction of urban solid waste (OFMSW) derivatives are useful as nutrients for MFCs. Developing countries generate 60% more waste than developed countries, according to research. Furthermore, two different types of OFMSW models (the one is fed air–cathode MFCs loaded with wastewater effluent and the other is cattle organic materials involving manure) have been identified for effective electricity generation. The highest power density in the manure-seeded (MFCs 123 mWm^2), wastewater stream inoculated (MFCs 123 mWm^2), and manure-seeded (MFCs 123 m (MFCs 116 mWm^2) was reported.

The phyla *Firmicutes* efficiency (67%) at the anode electrode has demonstrated their function in power generation by eliminating COD values (>86%) for all types of MFC designs, as well as in all mono- and polysaccharides compounds (>98%) (El-Chakhtoura et al., 2014). The removal of organic pollutants from OFMWS with landfilling techniques has resulted in the formation of energy that is sustainable, clean, and renewable. These can be accomplished using a tubular MFC by studying the impact of temperatures (ranging 20 to 30°C with 5°C increases) on bioreactor features and various wastes abatement. At 100 X external resistors, the maximum current stability has been increased from 197.7 to 344.4 mA/m^2, and the most elevated power density production (from 14.8 to 47.6 mW/m^2) has been tripled. Bacterial strains such as *Geobacter* have contributed to the fermentation process by inflowing function electrons to anode and cathode electrodes. *Bacteroides* and *Clostridium* sp. have also aided the fermentation process (Karluvali et al., 2015).

Plant-associated MFCs with soil-blended compost organic substances have been reported to produce sustainable energy (bioelectricity) by using rice crop plants. Rice crops are contained in five containers, with soil in the sixth, and

MFCs have manure with high voltage difference and power uniformity over time due to external resistance (100 Ω). In rice crop plants with 1% organic compost mixed soil, the maximum voltage value (700 mV) has been demonstrated. It has also been found that rice crops without organic compost have a 95% lower power density. The power density of paddy PMFCs with organic compost compounds was reported to be three times higher, indicating that the nature of the organic compost impacts power generation (Moqsud et al., 2015). The plant's *Spartina anglica* and *Arundinella anomala* have been investigated for plant-associated MFCs' concurrent biomass, bioelectricity generation, and the highest power production. *S. anglica* (16%) and *A. anomala* (8%), respectively, have been observed to produce average power in PMFC over periods of 7 to 13 weeks. Also, the P-MFC with the *Arundo donax* plant has generated no electricity. In a PMFC membrane surface area with *S. anglica*, the maximum power density obtained was 222 mW/m^2. Due to the availability of more nutrient compounds and the soil types' anaerobic nature, further biomass generation is accepted with all kinds of PMFCs designs with a higher root to shoot ratio (Helder et al., 2010).

2.3.2 BIOFUEL GENERATION

Biofuels are solid, liquid, or gaseous fuels that are produced from sustainable natural resources. Microalgal biomass is used in the same methods as terrestrial biomass for energy generation. The quality and quantity of biomass feedstock, the specific product required, and the preferred cost benefit from the material are all variables that affect the conversion process selection (Brennan & Owende, 2010). Biofuels made from microalgae have recently established much interest because they have many potentials to replace fossil fuel–based energy. Microalgal biomass produces numerous biofuels (biodiesel, bioethanol, and biohydrogen fuels), making it a sustainable green energy source. Microalgal biofuel production or conversion systems include harvesting, cultivation, and extraction procedures. Microalgae species with high photosynthesis capability and biomass synthesis can efficiently reduce the amount of carbon dioxide released into the environment, thus reducing global warming. Microalgae biomass can be grown quickly and has a higher synthesis capacity in salty water and non-arable or barren land. Microalgae with a fast growth can produce around 70% lipid contents in their cell, based on the species. They have also demonstrated the ability to survive in harsh environments in limited biomass synthesis. Microalgae's ability to generate electricity will also include crude lipid for transport fuel production (around 80%) as standard total energy content. Biofuels made from microalgae biomass include biodiesel, bio-oil, biofuel and gases, H$_2$ biofuel, methane, and various alcoholic fuels. Biofuel generation involving microalgae species can be an economically feasible method at particular scales of bioprocess advancement, with environmental and economic benefits (Bhagea et al., 2019). Several attempts have been made to commercialize microalgae species-derived biofuels with the participation of both government and private sector bodies. They were using traditional and more detailed techniques for algal biomass cultivation and harvesting, as well as the existing advancements in biofuel generation procedures (Tan et al., 2015).

Microalgae-derived biofuel generation can be obtained by estimating future economically sustainable methods and environmentally friendly nature through energy and carbon balancing with ecological effects on overall product costs. This could develop a complete energy balance through technological advances or intellectuals combined with highly optimized manufacturing operations. Pumping, building, drying, and dewatering are all used for it. The importance of a water management system for the environment, as well as the managing of carbon dioxide gas levels with appropriate nutrient supplementations, are essential restrict variables for the optimal microalgal biofuel process design and the scale-up choices. Carbon dioxide, water, and nutrient usage price reductions of up to 50% have been found at a minimal cost (Slade & Bauen, 2013).

Selecting lipid-rich microalgae poses a significant challenge for microalgal biodiesel research, due to the limited storage of lipid compounds or output in microalgal mass harvesting. Under optimal temperature and natural solar irradiation, *Graesiella* sp. WBG-1 accumulates a considerable amount of storage lipids (Wen et al., 2016). Plants/waste materials from palm oil, rapeseed, soybean, and sugarcane crops can be used to produce other biofuels.

2.3.3 BIOGASES SYNTHESIS

Methane is a component of biogas (an energy-rich gas), primarily formed by the biodegradation of organic matter in the absence of O_2 with various microbial strains. It is made from nutrient-dense waste, and its by-products (compost) are used as fertilizers. Multiple substances can affect methane gas production, depending on the substrate characteristics and reactor systems. Methane production is a complicated bioprocess involving a variety of microorganisms, and the quantity of energy required is determined by the development profile of the microbial population (also used in pre-treatment, saccharification, and fermentation) as well as process conditions (temperature) (Schnürer, 2016). Crops such as wheat straw are a low-cost, plentiful raw material for biogas generation. Its complex structure makes cellulose-degrading enzymes of microbial origin challenging to access, resulting in a slow degrading process. The decomposition rate is determined by the efficiency with which the microbial community system's enzymes affect the biogas (ammonia) process. In batch cultivation, the degradation of cellulose, wheat, and rice straw substrates is initiated with bacterial inoculum for the co-digestion (CD) phase of biogas wastewater treatment plants (WWTP). Two cel5 T-RFs have been linked to maximum degradation efficacy for both (rice and wheat) straw and cellulose. One of the corresponding cel5 partial genomic sequences has been found to share 100% identity with *Clostridium cellulolyticum* (Sun et al., 2016).

Anaerobic digestion has been inhibited and failed due to high salinity. The effect of increasing NaCl concentration on biogas synthesis was studied using Illumina high-throughput sequencing technologies. It was observed to be based on changes in the bacterial populations looking for higher salt concentrations. Due to no microbial inhibition in acidogenesis, NaCl concentrations of 20 gL^{-1} in the blank category containing *Clostridium, Bacteroides*, and uncultured BA021 showed an intense VFA concentration and specific CO_2 synthesizing rate. The H category containing bacterial populations such as *Soehngenia, Thermovirga*, and *Actinomyces* had a higher methane-producing capacity than the blank category, which resulted in a lower level of CH_4 production (42.2%) as well as a lower CH_4 production rate (37.12%) and pH. Bacterial populations were divided into the blank and H categories using Illumina sequencing to investigate hydrolytic and acidogenic capacities. Both categories include archaeal sp. such as *Methanolinea, Methanosaeta, Methanoculleus*, and *Methanospirillum*. Compared to acetoclastic methanogens, hydrogenotrophic methanogens are less resistant to high salinity (Wang et al., 2017).

Renewable energy production can address the world's central problem of energy scarcity, and biogas (methane) is one of the most promising renewable natural energy carriers. Its manufacture combines the removal of organic pollutants with the creation of a flexible energy source. The nature of the bacterial population with their metabolic processes participate in the biogas generation. Metagenomic methods are a new method evaluated to see if they can help resolve the operational and taxonomic ambiguity of bacterial consortium. Driving forces for optimum biogas synthesizing bacterial populations can reveal microbiological heterogeneity and its regulatory function in hydrogen metabolism (Campanaro et al., 2016). The bacterial population's rational design will result in a more efficient biogas synthesis process in large-scale applications. The SOLiDTM short-read DNA sequencing technology can help decipher systemic and functional contexts in the microbial community for optimal biogas-synthesized capability. *Clostridia*, a eubacterium, has been found to decompose organic matter. Archaea are a minority in this community, but they play a crucial role in biogas production. The most common species is *Methanoculleus marisnigri*, which is classified as a hydrogenotrophic methanogen. *Clostridia* contribute to the decomposition of organic substrates, as noted with its significance in the metabolism of hydrogen gas biofuel within biogas-producing bacterial populations (Wirth et al., 2012).

2.4   MICROALGAE FOR BIOENERGY PRODUCTION

Microalgae are photosynthetic prokaryotic or eukaryotic organisms that can create a network or community of a few micrometers to a hundred micrometers in size and are found worldwide. Algae are a taxonomic group with no well-defined classification (WEHR, 2007). They are terrestrial plants such as thallophytes with no branches, leaves, roots, protection around the cells, and rely on chlorophyll for photosynthetic pigment. Algae are classified into macroalgae that are multicellular and can grow to many meters in length. Microalgae are tiny organisms, varying in size from 0.2 m to 100 m or even more (Markou et al., 2012). Compared to conventional crops such as corn and soybean, microalgae have high areal biomass output, and the oil level in microalgae can surpass 80% of its dry weight biomass (Chisti, 2007). Blue-green algae (*Cyanophyceae*), red algae (*Rhodophyceae*), green algae (*Chlorophyceae*), brown algae (*Phaeophyceae*), and diatoms (*Bacillariophyceae*) are the five major groups of algae. Even though *Cyanobacteria* (blue-green algae) belong to the bacterial community, they are often referred to as "algae" because they are photosynthetic prokaryotes (Brodie & Lewis, 2009). Microalgae have several favorable characteristics that make them attractive for biofuel generation using the carbon source for algal growth in atmospheric carbon dioxide (Schenk et al., 2008). Many microalgae species thrive in non-potable water (brackish, wastewater, and seawater), and biofuel generation could be merged with either of these technologies. This combination doesn't compete with agriculture for arable land, and it certainly doesn't use freshwater resources. Algal biofuel production can be combined with CO_2 mitigation from flue gas, wastewater treatment, and high-value chemical productions (Benemann et al., 2003; Pandey et al., 2016). Many microalgae sp. contain large amounts of lipids that can be transesterified into biodiesel (Figure 2.1). Density, cold flow, flash point, viscosity, and calorific value are all characteristics shared by microalgal biodiesel and petroleum-based diesel.

Microalgae can be extracted in batches throughout the year, ensuring a consistent oil supply (Schenk et al., 2008). Unlike terrestrial plants, which require pesticides or herbicides, microalgae do not require chemicals, which are harmful to the environment and increase production costs. Microalgae and other biopolymers (observed in woody biomass) lack lignin, making biomass processing and conversion more complex (Alvira et al., 2010). Aside from that, residual algal biomass, mainly made up of proteins and carbohydrates, can be transformed into several biofuels, including methane and alcohol fuels. They could also be converted to other non-fuel co-products that can be processed and manufactured into high-value products like nutriceuticals, therapeutics, and animal feeds.

2.5 WASTEWATER AS AN ALTERNATE SOURCE OF NUTRITION FOR MICROALGAE HARVESTING

Microalgae multiply and can undergo one/two cell divisions a day in proper conditions (Sharma et al., 2011). They generate biomass during the photosynthesis process, which can be represented by the given formula:

$$\text{Light} + H_2O + CO_2 + \text{Nutrients} \Rightarrow \text{Biomass} + O_2 \qquad (2.1)$$

Microalgae harvesting involves a nutrient and salt-rich culture medium that allows them to multiply. Algal growth is influenced by various physicochemical and biological variables, including light, pH, temperature, and nutrient levels (Mata et al., 2010). The photosynthetic period has a significant impact on microalgae growth. When the photosynthetic time is extended from 6 to 12 h, the biomass concentration can reach 180% (Ip et al., 1982). Microalgae receive a large amount of inorganic carbon from the photosynthetic process, which, when associated with solar light, releases glucose and could be used as a carbon resource by the microalgal sp. as per the given equation:

$$6H_2O + 6CO_2 + Light \Rightarrow C_6H_{12}O_6 + 6O_2 \qquad (2.2)$$

Temperature is another significant variable that influences algal growth. Temperature affects the amount of biomass generated, especially in the first 7 days (Ip et al., 1982). Microalgae grow best at temperatures between 20 to 30°C. If these values are surpassed, the cells can be destroyed (Chisti, 2008). Depending on the microalgae type, high CO_2 concentrations may increase biomass production (Singh & Singh, 2014). Since inorganic carbon in carbonate form causes high pH values, the volume of CO_2 added to water is inversely related to the pH of the solution.

Microalgae harvesting also involves an aeration process that includes the CO_2 needed for photosynthesis and pH stabilization. The reactor must be slowly stirred to indicate that cells and nutrients are uniformly dispersed (Creswell, 2010). Microalgae are usually grown in photobioreactors either in open systems (raceway ponds, and tanks, open ponds, turf scrubbers) or closed systems (flat panels, coil systems, and tubular photobioreactors). The closed system enables more precise environmental monitoring and is more efficient at regulating growth factors. Thus, the input of CO_2 and specific growth are more effective. Open systems, on the other hand, can be more effective by using wastewater, and several microalgae species cultivated in effluents have minimal energy costs (Chiu et al., 2008; Muñoz et al., 2009). Microalgae provide a method that incorporates CO_2 recovery, wastewater remediation, and biofuel generation, which is ideal in considering the need for sustainable energy and the increasing demand for effective wastewater treatment. Increased levels of nutrient (phosphorus and nitrogen) removal have been reported in turf scrubber systems. This trend was found in the biomass preserved in three rivers in the Chesapeake Bay, USA, using a prototype turf scrubber device. The time of year was critical for remediation of harmful by-products in river water, with the effective outcomes showing elimination of up to 55% of total nitrogen and 65% of total phosphorus, which was both maintained in the biomass (Mulbry et al., 2010).

Compared to other systems like tanks and photobioreactors, the algae turf scrubber is a viable option for wastewater treatment. The turf scrubber system has several advantages, including temperature regulation in areas with significant solar incidents and the growth of a microbial culture that promotes nutrient removal through the use of microalgae, other microorganisms (bacteria and fungi). It is essential to collect biomass with the capability to produce biofuels under such conditions. However, ample oil levels in the biomass are a critical requirement for developing other biofuels, including bio-oil, bioethanol, and biogas, and others, that would enable the biomass to be fully exploited.

Since the oil content of biomass generated in closed systems is higher than that of biomass production in open systems, photobioreactors could be designed to manufacture feedstock for biofuels, including biodiesel and bioethanol. The photosynthetic performance of *Spirulina platensis* cultivated in photobioreactors was favorable, according to Ai et al. (Ai et al., 2008). The diatom *Chaetoceros calcitrans* multiplied in photobioreactors, with the highest specific growth rates (μ) of 9.65×10^{-2} h^{-1} and 8.88×10^6 cells mL^{-1} in batch and semicontinuous systems, respectively. The results for biomass production were better, with reduced light intensity (Krichnavaruk et al., 2005). The microalgae *Chlorella* sp. developed a satisfactory amount of biomass (1.445 ± 0.015 g L^{-1} of dry cells) when grown in a semicontinuous photobioreactor. CO_2 assimilation was improved by the growth, efficiency, and quantity of CO_2 eliminated acquired under conditions with high culture control and an increased inoculum concentration using cells that were already applied to the method (Chiu et al., 2008). The concentration of microalgae impacts the growth rate before it reaches an optimal level under operating conditions (Vasumathi et al., 2012).

As a result, microalgae can generate 3 to 10 times higher energy per hectare than other land cultures, and they're related to CO_2 reduction and wastewater treatment (Demirbas, 2010). Microalgae cultivation is a possible strategy for land plants to reduce their ecological effects. However, optimization of various system variables essential to the process efficacy, like lipid production, must be assumed (Kalt & Kranzl, 2011). A significant level of water pollution is created globally because of the growing world population and people's sophisticated living conditions. *Wastewater* is a term used to describe the end product of agricultural, domestic, industrial, and municipal sources (Bhatt et al., 2014). The wastewater quality reflects the lifestyles and technologies of the producing society. Organic matter such as proteins, carbohydrates, lipids, and volatile acids are found in wastewater, as well as inorganic matter such as calcium, sodium, sulfur, magnesium, chlorine, phosphate, ammonium salts, potassium, bicarbonate, and heavy metals (Abdel-Raouf et al., 2012). Eutrophication or algal blooms are caused by an accumulation of such nutrient levels in the aquatic environment caused by anthropogenic waste generation. Every year, the European Union produces over 300 million tons of degradable domestic contaminants, industrial effluents, and other wastes, with the vast majority of these wastes going unused (Mccormick & Kautto, 2013). Humans generate nearly 3 billion tons of household wastewater per year (Howard, 1933).

Annual population growth in India is expected to reach 600 million by 2030, putting a strain on urban flow back (wastewater), accounting for 70%–80% of the waterways (Amerasinghe et al., 2013). According to initial studies from the Central Pollution Control Board (CPCB) (CPCB, 2017), in New Delhi, India, the country's wastewater production is about 40 billion liters/day, primarily from large cities. Around 20%–30% of the produced wastewater is treated. Many microalgae sp. can survive in wastewater environments due to their potential to use excessive organic C as well as inorganic N and P (Pittman et al., 2011). Algae absorb these nutrients, as well as CO_2, and use them to obtain biomass through photosynthesis. Microalgae are the most common microorganisms used in oxidation ponds and oxidation ditches to treat domestic wastewater. Algae has also been used to treat wastewater at a low cost and environmentally friendly manner (Green et al., 1995; Kshirsagar, 2013; Paddock, 2019). The concept of using wastewater as a source for algae biofuel generation is familiar, as it was proposed in a study by the Aquatic Species Program (ASP) in the United States of America from 1978 to 1996 (Sheehan et al., 1978). The main challenge in developing a wastewater-based algae biofuel generation technology is identifying optimal microalgae species that can grow in a wastewater environment while removing large amounts of nutrients and producing high biomass and lipid productivity. Researchers worldwide have extensively investigated the viability of utilizing algae for biofuel production from wastewater with nutrient reduction properties, primarily N and P from effluents (Abou-Shanab et al., 2013; Bhatnagar et al., 2011; Hernández et al., 2013; Prathima Devi et al., 2012; Xin et al., 2010; Zhou et al., 2011). Industrial effluents, such as chemical and tannery wastewater, contain several metal ions with organic N and P components (Zhen-Feng et al., 2011), and heavy metal pollution makes it more toxic, which inhibits algal growth. When algae are cultivated in domestic wastewater, the total biofuel potential is $\sim$0.16 Mt/year (Chanakya et al., 2012), assuming a 20% lipid fraction.

2.5.1 Pre-Treatment of Industrial Wastewaters: Types and Salient Features

Prior to its use as the substrate for the cultivation of microalgae, the wastewater needs to be cleansed from various toxic substances that cause an adverse effect on the overall quality of the water. Pre-treatment reduces the following important Physico-chemical properties of the wastewater such as total suspended solids (TSS), total organic content (TOC), chemical oxygen demand (COD), volatile organic compounds (VOCs), oil and grease, etc., prior to its disposal, reuse, recycle, and reclaim. Over the years, a wide variety of techniques (both physical as well as chemical) have been employed for the pre-treatment of waste materials. One of the most widely used techniques in this regard is the thermal pre-treatment of the waste that increases both biodegradability and reusability of the material significantly. Thermal pre-treatment even at a temperature below 100°C on waste-activated sludge was reported to increase the biomethane production by 31% as compared to its untreated counterpart (Ruffino et al., 2015). Another widely employed pre-treatment method is ultrasonication that involves cavitation and production of hydroxyl radicals for the treatment. It is relatively cost-effective in nature as compared to its counterpart, ozonation (Sri Bala Kameswari et al., 2011). Another widely applied pre-treatment technology is the biological approach that involves enzymatic hydrolysis of the waste matter which cleaves the bonds of the specific substrates. This technique increases the biodegradability as well as the biomethane production rates while concomitantly decreasing the required processing time thus increasing the overall efficacy

(Meng et al., 2017; Valladão et al., 2007). Chemical approaches for the pre-treatment of the waste have also been widely employed during the remediation of a variety of organic pollutants. The addition of coagulants or a combination of coagulants such as polyacrylamide and polyaluminum chloride to the wastewater increases the floc size that enhances the flux of the permeate and inhibiting the fouling in the membrane filtration (Pan et al., 2005). Other coagulants such as aluminium sulphate and aluminium hydroxide were found to have superior fouling inhibition properties (Stoller, 2009). Another technique involves the use of strong reducing agents such as zero-valent iron (ZVI) that has been estab-lished to remove/reduce the organic pollutants from wastewater concomitantly increasing the biological wastewater treatment efficiency (J. W. Lee et al., 2009). In combination with other treatment techniques such as Fenton oxidation, it was found that the technique is able to reduce around 77% of the parent organic contaminant from the wastewater into non-toxic intermediates (Shen et al., 2013). The use of ozone for the pre-treatment of the waste is another commonly used method whose performance depends on the formation of hydroxyl radicals that alters the unsaturated bonds. Ozonation was found to increase the biode-gradability of the wastewater by 33% without causing any potential threat to the biomass (Battimelli et al., 2010). Wet air oxidation is another technique that oxidises the organic compounds into carbon dioxide and water under high pressure and temperature. The efficacy of this method can be increased by using different oxidants and catalysts. Photo-Fenton has been proved to be a promising alternative pre-treatment method for industrial application of waste. As com-pared to its ozone-based counterparts, it was reported to remove up to 76.9% of the initial COD concentration and 53.3% of dissolved organic carbon efficiently (Guzmán et al., 2016).

With extensive industrialization, the variety and the load of the effluent produced from the industries have magnified enormously. Considering this fact, a combined approach featuring different pre-treatment methods has been in-vestigated. The combination of AOPs has proved better results in terms of the reusability of the wastewater. Studies combining AOPs such as photo-Fenton and nano filtration for the pre-treatment of cork boiling wastewater yield high-quality effluents which could be re-used again in the process (Ponce-Robles et al., 2017). Similarly, a combination of Fenton oxidation with processes like flocculation and sedimentation resulted in higher COD removal efficiency up to as high as 94% and the parent pollutant phenol completely. It was also reported that the use of the adsorption technique along the above-mentioned technique lowers the iron concentration of the wastewater and thus minimizes the harmful effect on the membranes (Ochando-Pulido et al., 2012). Apparently, the results obtained from these studies proved the beneficial effect of the pre-treatment method on the overall wastewater characteristics. However, the salient features of these technologies have been discussed in the next section.

Pre-treatment of the industrial effluents affects the overall characteristics in various ways which have been discussed in the following section.

2.5.1.1   Enhancement in the Biodegradability of the Effluent

The most widely used practice for the remediation of several pollutants present in the industrial effluent is its biotreatment technology. However, the wastewater needs to have certain physicochemical and biological characteristics that enable it for treatment. Studies have suggested that the effluent must have a minimum biodegradability index of 0.4 to be treated biologically (Chamarro et al., 2001). Thus, a variety of pre-treatment methods such as catalytic ozonation, chemical oxidation, photocatalysis, Fenton oxidation, etc. have been implemented for the degradation of various contaminants found in the industrial effluents (X. Li et al., 2017; Mantzavinos & Psillakis, 2004; Perey et al., 2002). Enzymatic pre-treatment of the industrial wastewater results in an increase in biodegradability of the industrial effluent by more than twofold since the microorganisms are able to degrade the hydrolysed product more efficiently. Hydrolysis of the effluent from alcohol distillery by cellulase followed by aerobic biotreatment resulted in an increase in the oxidation rate by 2.3-fold thus establishing the significance of the pre-treatment technology (Sangave & Pandit, 2006). Another important pre-treatment technology that has displayed promising results is the Fenton method. Studies have shown that under optimum conditions the biodegradability of the industrial effluents having high COD values like that of pesticide industries increases exceedingly well (Chen et al., 2007). This is so because Fenton oxidation degrades phorate resulting in the mineralization of the intermediates into various non-toxic ions that can be utilized by the microbes as its nutrients. There are several studies that affirm the enhancement of the biodegradability of wastewater by electrochemical techniques (Prabakar et al., 2018). However, there are several factors such as pH, the distance between the electrodes, current density, ratio between the electrode area and the wastewater volume that plays a very significant role in the same. Studies reported a significant increase in the biodegradability of the contaminants from the wastewater of wet-spun acrylic fibers manufacturing industries by optimizing the above parameters (Gong et al., 2014). In recent times, a combination of various pre-treatment technologies has generated satisfying results. For example, a combination of coagulation-flocculation along with microscale ozonation has resulted in a significant increase in the biodegradability of the automobile coating wastewater (Xiong et al., 2017). Thus, it can be concluded that a different approach is necessary for the treatment of wastewater generated from different industries and to determine the most appropriate of them all, detailed characterization of the effluent is highly necessary.

2.5.1.2   Enhanced Production of Renewable Energy

In order to overcome the energy crisis, alternative sources of energy are being investigated. Nevertheless, the availability of the raw material is an additional cost for the same. Hence, the integration of waste management to the above process may solve the exigency. However, the industrial wastewater is of high organic load and comprises various inhibitor compounds that repress the biological activity and limits the complete utilization of the waste material for bioenergy production (Ennouri et al., 2016). Thus, pre-treatment of the waste or wastewater as the substrates is viewed as the solution to overcome this because this results in a higher yield of bioenergy production in a shorter time (Sivagurunathan et al., 2017; Veluchamy & Kalamdhad, 2017).

Owing to its versatile characteristics, bioethanol has proved itself to be a promising alternative source of bioenergy. The use of industrial waste which is a rich source of organic matter will prove to be an advantageous approach for the production of this non-renewable energy source rather than depending only on the food-based substrates (Y. Sun & Cheng, 2005). There are several studies pertaining to the use of industrial waste for the production of bioethanol. It has been observed that even minor pre-treatments such as pH adjustment, sterilization, or addition of magnesium ions enhance ethanol production considerably. This may be due to the fact that with sterilization the indigenous microorganisms were being killed and allows the growth of the required microorganism at optimum growth conditions (He et al., 2014). In another study, the waste sludge produced from anaerobic digestion was subjected to mechanical, chemical, thermal, or thermo-chemical pre-treatment methods and it was found that the pre-treatment increases the ethanol yield by 43% as compared to its untreated counterpart (Bashiri et al., 2016). The study also demonstrated the potential of alkaline pre-treatment for industrial-scale bioethanol production. However, from the practical point of view, both the cost and energy requirement of the process needs to be assessed. Hence, in order to develop a sustainable, cost-effective technology for bioethanol production, we have to evaluate the pre-treatment methods on a case-by-case basis.

Biohydrogen, known for its beneficial properties such as the absence of ozone harming emissions during burning, also known as clean-burning, has proved to be another source of sustainable power transporter (Boran et al., 2012; Sivagurunathan et al., 2017). Several studies have been carried out where different industrial effluents are subjected to various pre-treatment methods for the production of biohydrogen. One of those studies reported that heat shock of 100°C for 2 hours followed by acid pre-treatment of the effluent from chemical industries has proved to enhance the hydrogen production by a significant amount (Venkata Mohan et al., 2007). Augmenting it with glucose and sewage wastewater improves the yield up to 1.25 mmol of hydrogen per gram of COD evaluated. It has been evidenced that thermal pre-treatment of corn syrup effluent enhanced the hydrogen production rate from 10 L to 34 L per day (Hafez et al., 2009). In another study, heating the sludge at 80°C for 30 minutes has enhanced the production of hydrogen using wastewater from the beverage industry considerably (Kumar et al., 2015). Even production of biohydrogen employing third-generation technology using pre-treated wastewater has also yielded superior results. Chemically pre-treated sewage sludge used as the substrate for the production of biohydrogen by the marine microalgae yielded 78 ± 2.9 mL/0.05 g Volatile Solids, which establishes its efficiency as a promising alternative.

Studies have suggested that thermal pre-treatment of the wastewater during both primary and secondary treatment results in enhanced biogas production by 30–60% (Ennouri et al., 2016; Ferrer et al., 2008). Several other pre-treatment methods such as chemo-mechanical, microwave treatment, and ultrasonication have been studied using pulp mill waste sludge. Comparing the results demonstrated that microwave treatment is the most advantageous of all with an increase of more than 90% methane yield (Saha et al., 2011). Recent studies on ultrasonication as a pre-treatment method for waste-activated sludge yielded an increase of 31% of biogas production as compared to its untreated counterpart (Lizama et al., 2017). Co-pre-treatment of the waste material has also presented excellent results in terms of biogas production. Recent studies have evidenced that co-pre-treatment of food waste and waste-activated sludge resulted in an increase in methane production by more than 24% (J. Zhang et al., 2017).

The previous studies establish the fact that the use of organic waste for bioenergy production holds a sustainable future and these pre-treatment processes have the dual advantage of producing renewable bioenergy while simultaneously treating the wastewaters cost-effectively and sustainably.

2.5.2 Microalgae Biomass Production in Various Wastewaters

The commercialization efficiency of algal biofuels is heavily influenced by species selection, growth optimization, lipid content, and large-scale harvesting. The algal biomass generated and obtained by such wastewater treatment technology is being used to convert biofuels through varied mechanisms, including anaerobic digestion to biogas, carbohydrate fermentation to bioethanol, lipid transesterification to biodiesel, and high-temperature adaptation to crude oil (Craggs et al., 2011). Compared to conventional algal production by high-rate algal ponds (HRAPs), which consume freshwater and fertilizers, HRAPs can achieve the economic viability of algal biofuel generation from water treatment with low environmental impact (Park et al., 2011). The main challenge in microalgae study for existing HRAPs is to design an efficient and cost-effective carbonation system that can meet the high CO_2 demand while also improving biomass production (Putt et al., 2011).

Chlorella sp. was isolated from a wastewater pond and examined for lipid production efficiency in a bioreactor under photoautotrophic and heterotrophic conditions by Viswanath and Bux (Viswanath & Bux, 2012). The highest amount of biomass was obtained from *Chlorella* sp. cultivated under heterotrophic environmental conditions ($8.90~\text{gL}^{-1}$), relative to photoautotrophic (3.6-fold less biomass accumulation of high lipid content in cells), and autotrophic conditions (4.4-fold increasing lipid production). This research found that microalgae's heterotrophic growth is an effective method for producing biomass and higher lipid contents, lowering the price of producing algal biomass.

2.5.3 Advantages of Using Wastewater as an Alternate Source of Nutrition

Microalgae sp. can thrive in wastewater as they can use adequate organic C, as well as inorganic N and P. Even though the use of microalgae in treating wastewater have been widely supported, the traditional treatment approach is chemical waste extraction or activated sludge processing. The usage of microalgae in treating industrial wastewater is indeed limited; they are being used to treat wastewater on a small scale (Green et al., 1995; Pugazhendhi et al., 2020). The P precipitate will either be discarded in a landfill or processed further to create sludge fertilizer because the P obtained by these strategies is not correctly recyclable. Microalgae effectively remove nitrogen, phosphorus, and toxic heavy metals from wastewater (Ahluwalia & Goyal, 2007; Mallick, 2002) and thus have the capacity to enhance wastewater treatment, particularly during the last (tertiary) treatment process. Consequently, algae-based remediations are as effective as chemical treatments in removing P from wastewater (Pugazhendhi et al., 2020).

Algal technologies in wastewater treatment provide significant price reductions and reduced equipment than traditional chemical-based treatment approaches, making this solution more attractive to developing countries. For example, the substantial amount of O_2 generated by photosynthesizing microalgae would eliminate the requirement for mechanical aeration of the treatment pond and its high operational cost (Mallick, 2002). Treatment ponds must be oxygenated for heterotrophic aerobic bacteria to bioremediate organic and inorganic compounds effectively (Muñoz & Guieysse, 2006). Moreover, an algal technologies approach is more eco-friendly and effective because it produces reduced contaminants, including sludge by-products, and allows for effective nutrient recycling. Algal biomass, i.e. high in N and P, for example, could be used as a reduced fertilizer or animal feed (Muñoz & Guieysse, 2006; Wilkie & Mulbry, 2002). The majority of algal wastewater treatment research has accounted for lab-based small-scale, experimental high-rate algal ponds and pilot pond scale cultures.

Microalgae development in various wastewater conditions, primarily municipal sewage and agricultural manure wastewater, has been studied in several studies. These analyses have primarily been concerned with determining algae potential to eliminate N and P, as well as metals, from wastewater. These preliminary experiments, especially those that looked at factors for maximum algal biomass generation and strategies for cultivation from wastewater, would be instrumental in determining whether wastewater-grown microalgae can be used as a biofuel.

2.5.4 The Efficiency of Algal Growth in Wastewater

Several factors influence the efficient cultivation of microalgae in wastewater effluent. The pH, temperature, and the level of vital nutrients, including P, N, and organic C, as well as the presence of O_2, light, and CO_2, are all crucial factors. Microalgal growth in sewage water, for example, was shown to increase considerably during prolonged photoperiod conditions with CO_2 incorporation, whereas algal biomass decreased as temperature increased (Ip et al., 1982). The higher nutrient level of wastewater, including P and N, distinguishes wastewater from other growth media. Much of the nitrogen is in ammonia, which can inhibit algal growth at high concentrations (Ip et al., 1982; Konig et al., 1987; Wrigley & Toerien, 1990). Also, certain microbes in the wastewater may interact for vital nutrients with the microalgae. The initial density of algae in wastewater is also likely to determine population expansion (Lau et al., 1995). These factors will vary based on the wastewater source and the treatment facility.

Furthermore, the potential of various algal sp. to resist a specific wastewater environment will vary. Unicellular chlorophyte microalgae are highly tolerant of a wide range of wastewater environments and highly effective at absorbing nutrients (Aslan & Kapdan, 2006; González et al., 1997; Ruiz-Marin et al., 2010). *Chlorella* and *Scenedesmus* are the most dominant phytoplanktonic groups in oxidation ponds (Masseret et al., 2000) and high-rate algal ponds (Canovas et al., 1996). Nonetheless, the efficiency of different chlorophyte species varies. *Chlorella vulgaris* was considered more significant at absorbing P and N from wastewater than *Chlorella kessleri* in one study (Travieso et al., 1992). *Scenedesmus obliquus* performed better absorption in municipal wastewater than *C. vulgaris*.

2.5.5 ALGAL GROWTH IN MUNICIPAL SEWAGE WASTEWATER

Advanced municipal wastewater remediation includes an initial treatment process for solids sediment deposits, an intermediate treatment process for removing suspended and soluble organic compounds, and a final treatment process for maximum water treatment before discharge into the ecosystem. Several dissolved inorganic substances, such as P and N, are eliminated during this last treatment. Microalgae's ability to remove P and N through final wastewater treatment has been extensively investigated. Because specific unicellular green microalgae sp., particularly those belonging to the *Chlorella* and *Scenedesmus* genera, are remarkably tolerant of sewage effluent conditions, most studies have focused on their growth (Bhatnagar et al., 2009; Lau et al., 1995; Ruiz-Marin et al., 2010; Shi et al., 2007; Wang et al., 2010). Microalgae effectively extract nitrogen and phosphorus from municipal wastewater, whether in a free-swimming suspended or immobilized state. Numerous *Chlorella* and *Scenedesmus* sp., for example, can remove nitrate, ammonia, and total P from wastewater treatment with maximum (>80%) and, in certain conditions, nearly complete elimination, suggesting the capacity of microalgae for wastewater treatment (Martínez et al., 2000; Ruiz-Marin et al., 2010; Zhang et al., 2008). The microalgae exhibited a rapid growth rate during the batch growth phase in several of these experiments conducted in a laboratory environment. Ruiz-Marin et al. (2010) observed *S. obliquus* growth under a semicontinuous culture environment as well. They discovered that initial development was better in batch cultivation than in four cultivating phases (every 35 h with wastewater introduced at the beginning of each step) because of the subsequent nutrient reduction in the batch. The cell's proliferation and chlorophyll composition declined substantially after the fourth culture phase, indicating culture destruction.

Microalgae have also been shown to develop and extract nutrients from primarily settled municipal wastewater in studies. *C. vulgaris*, for example, has been shown to eliminate over 90% of N and 80% of P from treated wastewater effluent (Lau et al., 1995). In this analysis, the impact of modifying the initial algal inoculation density was compared with treatments varying from a concentrated inoculation of 1×10^7 cells mL^{-1} to a low-density inoculation of 5×10^5 cells mL^{-1}. It was found that the growth patterns of both treatments were not substantially different. But for the lower initial inoculum density, all procedures had the same total amount of nutrients eliminated. This indicates that successful wastewater and nutrient removal are unaffected by cell density at the outset. Two other recent studies looked at *Chlorella* sp. growth in raw wastewater. Wang et al. investigated *Chlorella* development in pre-treatment and from three different wastewater treatment stages. In wastewater before and after primary settling, many tested parameters, such as P and N elimination, metal ion reduction, and cell proliferation, were comparable. In the fourth phase of treated wastewater created by the sludge centrifuge, the algal development was significantly higher than in the other three treatments. The much higher P and N concentrations in this wastewater (131.5 mgL^{-1} total N and 201.5 mgL^{-1} total P) are most quick to attribute. The cells could grow well despite the non-optimal N:P ratio compared to traditional algal growth media (Li et al., 2008).

Chlorella minutissima, a *Chlorella* species found in wastewater treatment oxidation ponds in India, was described in a second latest report (Bhatnagar et al., 2009). *C. minutissima* could thrive in higher wastewater concentrations and govern the oxidation pond system's later stages. The biomass productivity of this algae was significantly elevated in mixotrophic (photoheterotrophic) environments, with 379 mgL^{-1} after 10 days compared to 73.03 mgL^{-1} in photoautotrophic conditions (Bhatnagar et al., 2009).

2.5.6 Algal Growth in Agricultural Wastewater

Agricultural wastewater, mainly generated from manure, has a higher P and N content than municipal wastewater (Wilkie & Mulbry, 2002). Despite the high nutrient concentrations, research has shown that microalgae can grow efficiently on agricultural waste. Microalgae effectively extract P and N from manure-based wastewater, just as they are in municipal wastewater (An et al., 2003; González et al., 1997; Wilkie & Mulbry, 2002). *Botryococcus braunii*, for example, developed well in piggery wastewater, including 788 mgL^{-1} NO_3, and eliminated 80% of the NO_3 (An et al., 2003). Since certain species of benthic freshwater algae have higher nutrient absorption rates than planktonic (suspended) algae, experiments on algal-mediated nutrient removal from dairy manure have evaluated the potential of intertidal freshwater algae instead of planktonic (suspended) algae (W. W. Mulbry & Wilkie, 2001; W. Mulbry et al., 2008; Wilkie & Mulbry, 2002). *Microspora willeana, Ulothrix* sp., and *Rhizoclonium hierglyphicum* are among these organisms. Algal development conditions and nutrient absorption were high, comparable to algae values grown in a semi-continuous harvesting system on municipal wastewater. The intertidal algae were grown in recycled wastewater with regular applications of fresh manure (Wilkie & Mulbry, 2002).

2.5.7 ALGAL GROWTH IN INDUSTRIAL WASTEWATER

There is a surge of attention in using algae to treat wastewaters, mainly to remove heavy metals (zinc, cadmium, chromium, etc.) and organic chemical contaminants (biocides, hydrocarbons, and surfactants) instead of P and N (Ahluwalia & Goyal, 2007; de-Bashan & Bashan, 2010; Mallick, 2002). Algal development rates are low in many industrial wastewaters due to reduced N and P concentrations and higher toxin concentrations. As a result, there is limited scope for using wastewaters for large-scale algal biomass production. Moreover, agricultural and industrial contaminants are more easily accessible and standardized, unlike the varied components of wastewaters. Similarly, an examination of carpet mill wastewater suggests that certain industrial wastewaters might potentially provide opportunities for substantial algal bio-mass production (Chinnasamy et al., 2010). Dalton, GA, USA, produces 100–115 million L of wastewater per day from carpet mill wastewater (along with a limited amount of municipal wastewater). Processing chemicals and coloring agents utilized in the factories, as well as a variety of inorganic elements, such as reduced metal concentrations and significantly lower total N and P concentrations, are all present in the wastewater (Chinnasamy et al., 2010). A significant quantity of biomass and probably biodiesel may be developed due to the enormous wastewater generated by this industry.

2.5.8 ALGAL GROWTH IN ARTIFICIAL WASTEWATER

Several findings looked at algal development and nutrient elimination properties using artificial wastewater (Aslan & Kapdan, 2006; K. Lee & Lee, 2001; Voltolina et al., 1999). The majority of synthetic wastewater media are made up of inorganic elements, such as significant amounts of particular nutrients, and are devoid of solid organic matter and other possible contaminants. As a result, utilizing artificial wastewater to measure real-world factors could have several disadvantages. Researchers discovered that, while nutrient removal rates in synthetic and municipal wastewater are similar, microalgal growth rates in ar-tificial wastewater are the highest (Lau et al., 1995; Ruiz-Marin et al., 2010). The explanations for this are likely to be enhanced contamination of real wastewaters, the inhibitory or competitive activity of indigenous microbes, and the waste-water's various chemical compositions.

2.6 MICROALGAE-BASED MICROBIAL FUEL CELL FOR BIOENERGY PRODUCTION

The microalgae-microbial fuel cell (mMFC) is a promising way to generate bioelectricity while also treating wastewater through a microbial process, enabling algae to treat a wide variety of contaminants at different concentrations (Lee et al., 2015). Several findings have shown that microalgal biomass can be used as a biocathode in mMFCs, either in suspended or attached forms (Figure 2.2) (González Del Campo et al., 2013; Liu et al., 2013; Venkata Mohan et al., 2014). Recent studies have explored the potential of producing zero-carbon electricity utilizing microalgae cultivated in a cathode chamber as a substrate for the anodic chamber's biofilm (Cui et al., 2014; Kondaveeti et al., 2014; Rashid et al., 2013). It was recently announced that mMFC could be used as a bioremediation unit for contaminants (Jiang et al., 20212; E. E. Powell et al., 2011; Powell et al., 2011). The development of catalysts, advanced materials, and substrates has reduced the total expenditure of mMFCs while increasing their applicability and performance (Baicha et al., 2016).

Powell et al. investigated the use of *Chlorella vulgaris* as a biocathode (Powell et al., 2011). Algae was found to act as an electron acceptor, allowing it to cultivate utilizing CO_2 produced at the cathode. Optimal process parameters would be able to maintain the flow of electrons and generate electricity indefinitely. The main benefit of using microalgae in mMFC is that they can simultaneously operate as an electron donor and a donor and acceptor (both in anodic and cathodic chambers). Furthermore, the mMFC produces electricity in both dark and light environments. Microalgae in mMFCs, on the other hand, is

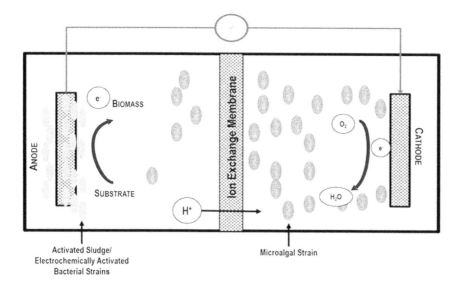

FIGURE 2.2   Schematic diagram of algae-based microbial fuel cells.

a relatively recent addition. Single chambers, double chambers, and photo-synthetic sediment are the three types of configurations. Seasonal sensitivity, optimal layout, a lack of data on bacterial symbiotic relationships, and light source constraints are all significant obstacles that seem to have been overcome (Cui et al., 2014; Saba et al., 2017).

Chlorella vulgaris has been proposed as an electron acceptor in the cathode chamber of MFCs as well as for CO_2 fixation (Chanakya et al., 2012). Wang et al. investigated the efficiency of a sediment MFC with a *Chlorella vulgaris* biocathode, achieving a cathode output of 21 mWm^2 (Park et al., 2011). Other, more complicated designs have also been made. Wu et al. for example constructed an MFC with a tubular photobioreactor system as the cathodic chamber and used *Chlorella vulgaris* to regenerate oxygen (Viswanath & Bux, 2012). Wang et al. developed a unique MFC configuration that allows microalgae to grow by releasing off-gases such as CO_2 into the cathode compartment. Microbial carbon capture cells (MCCs) form such systems that can generate a 5.6 Wm^3 in the anodic chamber. The cathode eliminated all of the CO_2 from the anode, while the system treated wastewater, thereby generated electricity. Cui et al. examined using dead microalgal biomass as an anode substrate and reusing CO_2 generated in the anode to develop microalgae in the cathode. With a Coulombic efficiency of 6.3%, this assembly produced the highest power density of 1.9 Wm^2 (Green et al., 1995). Other researchers have investigated the impact of cathode materials on MFC output in systems with algae-assisted cathodes. Kakarla and Min evaluated the biocathode materials of a C-fiber comb and blank C-paper. The findings revealed that a carbon fiber brush cathode has the highest biomass yield and power density (30 mWm^2 cathodes) (Mallick, 2002). The first stack-biotic-MFCs with photosynthesizing microalgae in the cathode was developed by Gajda et al. (Ruiz-Marin et al., 2010). Microalgae contain enough oxygen to meet the needs of oxygen reduction. The light intensity is another significant factor affecting algae-assisted cathodes. Wu et al. aimed at how varying light intensities affected photo-microbial fuel cells (photo MFCs) with a *Desmodesmus* sp. A8 biocathode (Masseret et al., 2000). They discovered that variations in light intensities, and thus the voltage, had a significant impact on the anode and cathode resistances. When the light intensity is about 3000 lx, the voltage produced achieves a maximum, according to the findings. Another study examined how the continuous flow method affected the efficiency of *Chlorella vulgaris*–assisted cathode MFCs. In constant mode, Gonzalez del Campo et al. obtained a maximum energy production than in the sequencing-batch method (Ahluwalia & Goyal, 2007).

2.7 CONCLUSION

The present study discusses the various technologies used for bioenergy production and their advantages and limitations. Microalgae appear to have the potential to be used as a low-cost biomaterial for the production of sustainable energy. However, there exists a few limitations such as the need for pre-treatment, complex methods for oil extractions in the processes that can generate these green energies. The use of microalgae in the MFCs is one of the most promising applications that fix carbon dioxide and simultaneously treats wastewater producing high value-added products alongwith. Microalgae can either be used as a substrate and anode or can be used as a cathode that produces oxygen and fixes carbon dioxide. However, further studies need to be carried out to maximize the overall performance of MFCs using microalgae as feedstock. There are many factors such as correct strain selection, optimum experimental conditions, and better substrate adoption are a few to name that affect the bioenergy production using microalgae-based MFC. The application of genetic engineering will allow the development of effective microbial candidates that can yield higher bioenergy with reduced experimental time and efficient use of different organic waste material in the effluents that benefits both the industry as well as the environment.

REFERENCES

Abdel-Raouf, N., Al-Homaidan, A. A., & Ibraheem, I. B. M. (2012). Microalgae and wastewater treatment. *Saudi Journal of Biological Sciences*, *19* (3), 257–275. 10.1016/j.sjbs.2012.04.005. Elsevier.

Abou-Shanab, R. A. I., Ji, M. K., Kim, H. C., Paeng, K. J., & Jeon, B. H. (2013). Microalgal species growing on piggery wastewater as a valuable candidate for nutrient removal and biodiesel production. *Journal of Environmental Management*, *115*, 257–264. 10.1016/j.jenvman.2012.11.022

Ahluwalia, S. S., & Goyal, D. (2007). Microbial and plant derived biomass for removal of heavy metals from wastewater. *Bioresource Technology*, 98 (12), 2243–2257. 10.1016/j.biortech.2005.12.006. Elsevier.

Ai, W., Guo, S., Qin, L., & Tang, Y. (2008). Development of a ground-based space microalgae photo-bioreactor. *Advances in Space Research*, *41*(5), 742–747. 10.1016/j.asr.2007.06.060

Alatraktchi, F. A. Z. a., Zhang, Y., & Angelidaki, I. (2014). Nanomodification of the electrodes in microbial fuel cell: Impact of nanoparticle density on electricity production and microbial community. *Applied Energy*, *116*, 216–222. 10.1016/j.apenergy.2013.11.058

Alvira, P., Tomás-Pejó, E., Ballesteros, M., & Negro, M. J. (2010). Pretreatment technologies for an efficient bioethanol production process based on enzymatic hydrolysis: A review. *Bioresource Technology*, *101*(13), 4851–4861. 10.1016/j.biortech.2009.11.093

Amerasinghe P., Bhardwaj R. M., Scott C., Jella K., & Marshall F. (2013). *Urban wastewater and agricultural reuse challenges in India*. IWMI Research Report 147, www.iwmi.org. https://www.gwp.org/globalassets/global/toolbox/references/urban-wastewater-and-agricultural-reuse.-challenges-in-india-iwmi-2013.pdf

An, J.-Y., Sim, S.-J., Suk Lee, J., & Woo Kim, B. (2003). *Hydrocarbon production from secondarily treated piggery wastewater by the green alga Botryococcus braunii.*

Asaoka, K., & Atsumi, M. (2007). *Seaweed bioethanol production in Japan-The Ocean Sunrise Project Masahito Aizawa, Vice Chairman Association of Quality Assurance for Marine Products Toshitsugu Sakou*. https://doi.org/10.1109/OCEANS.2007.4449162

Aslan, S., & Kapdan, I. K. (2006). Batch kinetics of nitrogen and phosphorus removal from synthetic wastewater by algae. *Ecological Engineering*, *28*(1), 64–70. 10.1016/j.ecoleng.2006.04.003

Baicha, Z., Salar-García, M. J., Ortiz-Martínez, V. M., Hernández-Fernández, F. J., de los Ríos, A. P., Labjar, N., Lotfi, E., & Elmahi, M. (2016). A critical review on microalgae as an alternative source for bioenergy production: A promising low cost substrate for microbial fuel cells. *Fuel Processing Technology* 154, 104–116. Elsevier B.V. 10.1016/j.fuproc.2016.08.017

Bashiri, R., Farhadian, M., Asadollahi, M. A., & Jeihanipour, A. (2016). Anaerobic digested sludge: A new supplementary nutrient source for ethanol production. *International Journal of Environmental Science and Technology, 13*(3), 763–772. 10.1007/s13762-015-0925-8

Battimelli, A., Loisel, D., Garcia-Bernet, D., Carrere, H., & Delgenes, J. P. (2010). Combined ozone pretreatment and biological processes for removal of colored and biorefractory compounds in wastewater from molasses fermentation industries. *Journal of Chemical Technology and Biotechnology, 85*(7), 968–975. 10.1002/jctb.2388

Benemann, J. R., van Olst, J. C., Massingill, M. J., Weissman, J. C., & Brune, D. E. (2003). *The controlled eutrophication process: Using microalgae for CO_2 utilization and agricultural fertilizer recycling*. In: Greenhouse Gas Control Technologies - 6th International Conference. pp. 1433–1438. https://doi.org/10.1016/b978-008044276-1/50227-0

Bhagea, R., Bhoyroo, V., & Puchooa, D. (2019). Microalgae: The next best alternative to fossil fuels after biomass. A review. *Microbiology Research, 10*(1), 7936. 10.4081/mr.2019.7936

Bhatnagar, A., Bhatnagar, M., Chinnasamy, S., & Das, K. C. (2009). *Chlorella minutissima – A promising fuel alga for cultivation in municipal wastewaters*. 10.1007/s12010-009-8771-0.

Bhatnagar, A., Chinnasamy, S., Singh, M., & Das, K. C. (2011). Renewable biomass production by mixotrophic algae in the presence of various carbon sources and wastewaters. *Applied Energy, 88*(10), 3425–3431. 10.1016/j.apenergy.2010.12.064

Bhatt, N. C., Panwar, A., Bisht, T. S., Tamta, S., Zhou, G., & Zhuang, L. (2014). Coupling of algal biofuel production with wastewater. *Scientific World Journal*. 10.1155/2014/210504.

Boran, E., Özgür, E., Yücel, M., Gündüz, U., & Eroglu, I. (2012). Biohydrogen production by *Rhodobacter capsulatus* in solar tubular photobioreactor on thick juice dark fermenter effluent. *Journal of Cleaner Production, 31*, 150–157. 10.1016/j.jclepro.2012.03.020

Brennan, L., & Owende, P. (2010). Biofuels from microalgae – A review of technologies for production, processing, and extractions of biofuels and co-products. *Renewable and Sustainable Energy Reviews*, 14(2), 557–577. Pergamon. 10.1016/j.rser.2009.10.009

Brodie J., & Lewis J. (2009). *Unravelling the algae: The past, present, and future of algal systematics* (Vol. 75). Systematics Association Special Volumes. CRC Press, London, UK. https://academic.oup.com/botlinnean/article/160/4/444/2418331

Calle, A., Montagnini, F., & Zuluaga, A. F. (2009). *Mata Atlantica-Michelin Ecological Reserve View project Organic Coffee in Agroforestry Experiments in Costa Rica View project*. https://www.researchgate.net/publication/284763306

Campanaro, S., Treu, L., Kougias, P. G., de Francisci, D., Valle, G., & Angelidaki, I. (2016). Metagenomic analysis and functional characterization of the biogas microbiome using high throughput shotgun sequencing and a novel binning strategy. *Biotechnology for Biofuels, 9*(1). 10.1186/s13068-016-0441-1

Canovas, S., Picot, B., Casellas, C., Zulkifi, H., Dubois, A., & Bontoux, J. (1996). Seasonal development of phytoplankton and zooplankton in a high-rate algal pond. *Water Science and Technology, 33*(7), 199–206. 10.1016/0273-1223(96)00355-1

Chamarro, E., Marco, A., & Esplugas, S. (2001). Use of Fenton reagent to improve organic chemical biodegradability. *Water Research*, *35*(4), 1047–1051. 10.1016/S0043-1354(00)00342-0

Chanakya, H. N., Mahapatra, D. M., Ravi, S., Chauhan, V. S., & Abitha, R. (2012). Reviews sustainability of large-scale algal biofuel production in India. *Journal of the Indian Institute of Science*, *92*, 63–98.

Chandrasekhar, K., Amulya, K., & Venkata Mohan, S. (2014). Solid phase bioelectrofermentation of food waste to harvest value-added products associated with waste remediation. *Waste Management*, *45*, 57–65.

Chandrasekhar, K., Kumar, A. N., Raj, T., Kumar, G., Kim, S.-H., 2021a. Bioelectrochemical system-mediated waste valorization. *Systems Microbiology and Biomanufacturing*, *1*, 1–12. https://doi.org/10.1007/S43393-021-00039-7

Chandrasekhar, K., Lee, Y. J., & Lee, D. W. (2015). Biohydrogen production: Strategies to improve process efficiency through microbial routes. *International Journal of Molecular Sciences*, *16*, 8266–8293.

Chandrasekhar, K., Mehrez, I., Kumar, G., Kim, S.-H. (2021b). Relative evaluation of acid, alkali, and hydrothermal pretreatment influence on biochemical methane potential of date biomass. *Journal of Environmental Chemical Engineering*, *9*, 106031. https://doi.org/10.1016/J.JECE.2021.106031

Chandrasekhar, K., Naresh Kumar, A., Kumar, G., Kim, D. H., Song, Y. C., & Kim, S. H. (2021c). Electro-fermentation for biofuels and biochemicals production: Current status and future directions. *Bioresource Technology*, *323*, 124598.

Chandrasekhar, K., & Venkata Mohan, S. (2012). Bio-electrochemical remediation of real field petroleum sludge as an electron donor with simultaneous power generation facilitates biotransformation of PAH: Effect of substrate concentration. *Bioresource Technology*, *110*, 517–525.

Chen, S., Sun, D., & Chung, J. S. (2007). Treatment of pesticide wastewater by moving-bed biofilm reactor combined with Fenton-coagulation pretreatment. *Journal of Hazardous Materials*, *144*(1–2), 577–584. 10.1016/j.jhazmat.2006.10.075

Chinnasamy, S., Bhatnagar, A., Hunt, R. W., & Das, K. C. (2010). Microalgae cultivation in a wastewater dominated by carpet mill effluents for biofuel applications. *Bioresource Technology*, *101*(9), 3097–3105. 10.1016/j.biortech.2009.12.026

Chisti, Y. (2007). Biodiesel from microalgae. *Biotechnology Advances*, *25*(3), 294–306. 10.1016/j.biotechadv.2007.02.001

Chisti, Y. (2008). Biodiesel from microalgae beats bioethanol. *Trends in Biotechnology*, *26*(3), 126–131. 10.1016/j.tibtech.2007.12.002

Chiu, S. Y., Kao, C. Y., Chen, C. H., Kuan, T. C., Ong, S. C., & Lin, C. S. (2008). Reduction of CO_2 by a high-density culture of Chlorella sp. in a semicontinuous photobioreactor. *Bioresource Technology*, *99*(9), 3389–3396. 10.1016/j.biortech.2007.08.013

CPCB. (2017). *Guidelines on environmental management of construction & demolition (C & D) wastes*. https://cpcb.nic.in/openpdffile.php?id=UmVwb3J0RmlsZXMvNT-UyXzE1MTEyNjQwMTVfbWVkaWFFwaG90bzQ2OTAucGRm

Craggs, R. J., Heubeck, S., Lundquist, T. J., & Benemann, J. R. (2011). Algal biofuels from wastewater treatment high rate algal ponds. *Water Science and Technology*, *63*(4), 660–665. 10.2166/wst.2011.100

Creswell L. R. (2010). Phytoplankton culture for aquaculture feed. Southern Regional Aquaculture Center, 5004.

Cui, Y., Rashid, N., Hu, N., Rehman, M. S. U., & Han, J. I. (2014). Electricity generation and microalgae cultivation in microbial fuel cell using microalgae-enriched anode and bio-cathode. *Energy Conversion and Management*, *79*, 674–680. 10.1016/j.enconman.2013.12.032

Cushion, E., Whiteman, A., & Dieterle, G. (2009). *Bioenergy development issues and impacts for poverty and natural resource management.* Agricultural and Rural Development Notes; No. 49. World Bank, Washington, DC. © World Bank. https://openknowledge.worldbank.org/handle/10986/9491 License: CC BY 3.0 IGO.

de-Bashan, L. E., & Bashan, Y. (2010). Immobilized microalgae for removing pollutants: Review of practical aspects. *Bioresource Technology*, *101*(6), 1611–1627. 10.1016/j.biortech.2009.09.043

Demirbas, A. (2009). Biofuels securing the planet's future energy needs. *Energy Conversion and Management*, *50*(9), 2239–2249. 10.1016/j.enconman.2009.05.010

Demirbas, A. (2010). Use of algae as biofuel sources. *Energy Conversion and Management*, *51*(12), 2738–2749. 10.1016/j.enconman.2010.06.010

Douskova, I., Doucha, J., Livansky, K., MacHat, J., Novak, P., Umysova, D., Zachleder, V., & Vitova, M. (2009). Simultaneous flue gas bioremediation and reduction of microalgal biomass production costs. *Applied Microbiology and Biotechnology*, *82*(1), 179–185. 10.1007/s00253-008-1811-9

Elbehri, A., Segerstedt, A., & Liu, P. (2012). *Biofuels and the sustainability challenge – A global assessment of sustainability issues, trends and policies for biofuels and related feedstocks.* FAO.

El-Chakhtoura, J., El-Fadel, M., Rao, H. A., Li, D., Ghanimeh, S., & Saikaly, P. E. (2014). Electricity generation and microbial community structure of air-cathode microbial fuel cells powered with the organic fraction of municipal solid waste and inoculated with different seeds. *Biomass and Bioenergy*, *67*, 24–31. 10.1016/j.biombioe.2014.04.020

ElMekawy, A., Srikanth, S., Bajracharya, S., Hegab, H. M., Nigam, P. S., Singh, A., Mohan, S. V., & Pant, D. (2015). Food and agricultural wastes as substrates for bioelectrochemical system (BES): The synchronized recovery of sustainable energy and waste treatment. *Food Research International*, *73*, 213–225. Elsevier Ltd. 10.1016/j.foodres.2014.11.045

Enamala, M. K., Pasumarthy, D. S., Gandrapu, P. K., Chavali, M., Mudumbai, H., Kuppam, C. (2019). Production of a variety of industrially significant products by biological sources through fermentation. In: Arora, P. K. (Ed.), *Microbial technology for the welfare of society* (pp. 201–221). Springer Singapore, Singapore. https://doi.org/10.1007/978-981-13-8844-6_9

Ennouri, H., Miladi, B., Diaz, S. Z., Güelfo, L. A. F., Solera, R., Hamdi, M., & Bouallagui, H. (2016). Effect of thermal pretreatment on the biogas production and microbial communities balance during anaerobic digestion of urban and industrial waste activated sludge. *Bioresource Technology*, *214*, 184–191. 10.1016/j.biortech.2016.04.076

Ferrer, I., Ponsá, S., Vázquez, F., & Font, X. (2008). Increasing biogas production by thermal (70°C) sludge pre-treatment prior to thermophilic anaerobic digestion. *Biochemical Engineering Journal*, *42*(2), 186–192. 10.1016/j.bej.2008.06.020

Gambino, E., Chandrasekhar, K., & Nastro, R. A. (2021). SMFC as a tool for the removal of hydrocarbons and metals in the marine environment: A concise research update. *Environmental Science and Pollution Research*, *28*, 1–16. https://doi.org/10.1007/s11356-021-13593-3

Gan, J., & Smith, C. T. (2012). Biomass utilization allocation in biofuel production: Model and application. *International Journal of Forest Engineering*, *23*(1), 38–47. 10.1080/14942119.2012.10739959

Gong, C., Zhang, Z., Li, H., Li, D., Wu, B., Sun, Y., & Cheng, Y. (2014). Electrocoagulation pretreatment of wet-spun acrylic fibers manufacturing wastewater to improve its biodegradability. *Journal of Hazardous Materials*, *274*, 465–472. 10.1016/j.jhazmat. 2014.04.033

González, L. E., Cañizares, R. O., & Baena, S. (1997). Efficiency of ammonia and phosphorus removal from a Colombian agroindustrial wastewater by the microalgae *Chlorella vulgaris* and *Scenedesmus dimorphus*. *Bioresource Technology*, *60*(3), 259–262. 10.1016/S0960-8524(97)00029-1

González Del Campo, A., Cañizares, P., Rodrigo, M. A., Fernández, F. J., & Lobato, J. (2013). Microbial fuel cell with an algae-assisted cathode: A preliminary assessment. *Journal of Power Sources*, *242*, 638–645. 10.1016/j.jpowsour.2013.05.110

Green, F. B., Lundquist, T. J., & Oswald, W. J. (1995). Energetics of advanced integrated wastewater pond systems. *Water Science and Technology*, *31*(12), 9–20. 10.1016/ 0273-1223(95)00488-9

Guzmán, J., Mosteo, R., Sarasa, J., Alba, J. A., & Ovelleiro, J. L. (2016). Evaluation of solar photo-Fenton and ozone based processes as citrus wastewater pre-treatments. *Separation and Purification Technology*, *164*, 155–162. 10.1016/j.seppur.201 6.03.025

Hafez, H., Nakhla, G., & el Naggar, H. (2009). Biological hydrogen production from corn-syrup waste using a novel system. *Energies*, *2*(2), 445–455. 10.3390/en202 00445

Hannon, M., Gimpel, J., Tran, M., Rasala, B., & Mayfield, S. (2010). Biofuels from algae: Challenges and potential. *Biofuels*, *1*, 763–784. https://doi.org/10.4155/bfs.10.44.

He, M.-X., Qin, H., Yin, X.-B., Ruan, Z.-Y., Tan, F.-R., Wu, B., Shui, Z.-X., Dai, L.-C., & Hu, Q.-C. (2014). Direct ethanol production from dextran industrial waste water by Zymomonas mobilis. *Korean J. Chem. Eng*, *31*(11), 2003–2007. 10.1007/s11 814-014-0108-1

Helder, M., Strik, D. P. B. T. B., Hamelers, H. V. M., Kuhn, A. J., Blok, C., & Buisman, C. J. N. (2010). Concurrent bio-electricity and biomass production in three Plant-Microbial Fuel Cells using Spartina anglica, Arundinella anomala and Arundo donax. *Bioresource Technology*, *101*(10), 3541–3547. 10.1016/j.biortech.2009.12.124

Hernández, D., Riaño, B., Coca, M., & García-González, M. C. (2013). Treatment of agro-industrial wastewater using microalgae-bacteria consortium combined with anaerobic digestion of the produced biomass. *Bioresource Technology*, *135*, 598–603. 10.1016/j.biortech.2012.09.029

Hill, J., Nelson, E., Tilman, D., Polasky, S., & Tiffany, D. (2006). Environmental, economic, and energetic costs and benefits of biodiesel and ethanol biofuels. *Proceedings of the National Academy of Sciences of the United States of America*, *103*(30), 11206–11210. 10.1073/pnas.0604600103

Howard, A. (1933). The waste products of agriculture: their utilization as humus. In: *Source: Journal of the Royal Society of Arts*, *82*(4229). http://www.jstor.org URL: http://www.jstor.org/stable/41360014

Htet, M. Z., Lim, Y., Ling, S., Hui, Y., & Rajee, O. (2013). Biofuel from microalgae-a review on the current status and future trends. In: *International Journal of Advanced Biotechnology and Research*, *4*. http://www.bipublication.com

IEA. (2013). *Tracking Clean Energy Progress 2013 – Analysis – IEA*. https://www.iea. org/reports/tracking-clean-energy-progress-2013

Ip, S. Y., Bridger, J. S., Chin, C. T., Martin, W. R. B., & Raper, W. G. C. (1982). Algal growth in primary settled sewage. The effects of five key variables. *Water Research*, *16*(5), 621–632. 10.1016/0043-1354(82)90083-5

Jia, J., Tang, Y., Liu, B., Wu, D., Ren, N., & Xing, D. (2013). Electricity generation from food wastes and microbial community structure in microbial fuel cells. *Bioresource Technology*, *144*, 94–99. 10.1016/j.biortech.2013.06.072

Jiang, H., Luo, S., Shi, X., Dai, M., & Guo, R.-B. (2012). A novel microbial fuel cell and photobioreactor system for continuous domestic wastewater treatment and bioelectricity generation. *Biotechnology Letters*, *34*, 1269–1274. 10.1007/s10529-012-0899-2.

Kabir, E., Kumar, P., Kumar, S., Adelodun, A. A., & Kim, K. H. (2018). Solar energy: Potential and future prospects. *Renewable and Sustainable Energy Reviews*, *82*, 894–900. 10.1016/j.rser.2017.09.094

Kadier, A., Chandrasekhar, K., & Kalil, M. S. (2017). Selection of the best barrier solutions for liquid displacement gas collecting metre to prevent gas solubility in microbial electrolysis cells. *International Journal of Renewable Energy Technology*, *8*, 93. https://doi.org/10.1504/IJRET.2017.086807

Kalt, G., & Kranzl, L. (2011). Assessing the economic efficiency of bioenergy technologies in climate mitigation and fossil fuel replacement in Austria using a techno-economic approach. *Applied Energy*, *88*(11), 3665–3684. 10.1016/j.apenergy.2011.03.014

Karluvali, A., Köroğlu, E. O., Manav, N., Çetinkaya, A. Y., & Özkaya, B. (2015). Electricity generation from organic fraction of municipal solid wastes in tubular microbial fuel cell. *Separation and Purification Technology*, *156*, 502–511. 10.1016/j.seppur.2015.10.042

Kojima, E., & Zhang, K. (1999). Growth and hydrocarbon production of microalga *Botryococcus braunii* in bubble column photobioreactors. *Journal of Bioscience and Bioengineering*, *87*(6), 811–815. 10.1016/S1389-1723(99)80158-3

Kondaveeti, S., Soon Choi, K., Kakarla, R., & Min, B. (2014). Microalgae Scenedesmus obliquus as renewable biomass feedstock for electricity generation in microbial fuel cells (MFCs). *Frontiers of Environmental Science & Engineering*, *8*, 784–791. 10.1007/s11783-013-0590-4.

Konig, A., Pearson, H. W., & Silva, S. A. (1987). Ammonia toxicity to algal growth in waste stabilization ponds. *Water Science and Technology*, *19*(12), 115–122. http://iwaponline.com/wst/article-pdf/19/12/115/98317/115.pdf

Krichnavaruk, S., Loataweesup, W., Powtongsook, S., & Pavasant, P. (2005). Optimal growth conditions and the cultivation of *Chaetoceros calcitrans* in airlift photobioreactor. *Chemical Engineering Journal*, *105*(3), 91–98. 10.1016/j.cej.2004.10.002

Kshirsagar, A. D. (2013). Bioremediation of wastewater by using microalgae: An experimental study. *International Journal of Life Sciences Biotechnology and Pharma Research*, *2*(3), 2250–3137. http://www.ijlbpr.com/currentissue.php

Kumar, G., Bakonyi, P., Sivagurunathan, P., Kim, S. H., Nemestóthy, N., Bélafi-Bakó, K., & Lin, C. Y. (2015). Enhanced biohydrogen production from beverage industrial wastewater using external nitrogen sources and bioaugmentation with facultative anaerobic strains. *Journal of Bioscience and Bioengineering*, *120*(2), 155–160. 10.1016/j.jbiosc.2014.12.011

Kumar, P., Chandrasekhar, K., Kumari, A., Sathiyamoorthi, E., & Kim, B. S. (2018). Electro-fermentation in aid of bioenergy and biopolymers. *Energies*, *11*(2), 343. doi: https://doi.org/10.3390/en11020343.

Lau, P. S., Tam, N. F. Y., & Wong, Y. S. (1995). Effect of algal density on nutrient removal from primary settled wastewater. *Environmental Pollution*, *89*(1), 59–66. 10.1016/0269-7491(94)00044-E

Lee, D. J., Chang, J. S., & Lai, J. Y. (2015). Microalgae-microbial fuel cell: A mini review. *Bioresource Technology*, *198*, 891–895. 10.1016/j.biortech.2015.09.061. Elsevier Ltd.

Lee, H. S., Parameswaran, P., Kato-Marcus, A., Torres, C. I., & Rittmann, B. E. (2008). Evaluation of energy-conversion efficiencies in microbial fuel cells (MFCs)

utilizing fermentable and non-fermentable substrates. *Water Research*, *42*(6–7), 1501–1510. 10.1016/j.watres.2007.10.036

Lee, J. W., Cha, D. K., Oh, Y. K., Ko, K. B., & Song, J. S. (2009). Zero-valent iron pretreatment for detoxifying iodine in liquid crystal display (LCD) manufacturing wastewater. *Journal of Hazardous Materials*, *164*(1), 67–72. 10.1016/j.jhazmat. 2008.07.147

Lee, K., & Lee, C.-G. (2001). Effect of light/dark cycles on wastewater treatments by microalgae. *Biotechnology and Bioprocess Engineering*, 6, 194–199.

Li, X., Guo, L., Sun, S., Zhang, X., Liu, G., Yao, H., Shi, M., Li, J., Yu, X., & Zhang, S. (2017). N-doped activated carbons derived from biological sludge, generated in petrochemical industries for supercapacitor applications. *Journal of Nanoscience and Nanotechnology*, *17*(9), 6655–6661. 10.1166/jnn.2017.14446

Li, Y., Horsman, M., Wang, B., Wu, N., & Lan, C. Q. (2008). Effects of nitrogen sources on cell growth and lipid accumulation of green alga Neochloris oleoabundans. *Applied Microbiology and Biotechnology*, *81*(4):629–636.10.1007/s00253-008-1681-1.

Liaquat, A. M., Masjuki, H. H., Kalam, M. A., Varman, M., Hazrat, M. A., Shahabuddin, M., & Mofijur, M. (2012). Application of blend fuels in a diesel engine. *Energy Procedia*, *14*, 1124–1133. 10.1016/j.egypro.2011.12.1065

Liu, X. W., Sun, X. F., Huang, Y. X., Li, D. B., Zeng, R. J., Xiong, L., Sheng, G. P., Li, W. W., Cheng, Y. Y., Wang, S. G., & Yu, H. Q. (2013). Photoautotrophic cathodic oxygen reduction catalyzed by a green alga, *Chlamydomonas reinhardtii*. *Biotechnology and Bioengineering*, *110*(1), 173–179. 10.1002/bit.24628

Lizama, A. C., Figueiras, C. C., Herrera, R. R., Pedreguera, A. Z., & Ruiz Espinoza, J. E. (2017). Effects of ultrasonic pretreatment on the solubilization and kinetic study of biogas production from anaerobic digestion of waste activated sludge. *International Biodeterioration and Biodegradation*, *123*, 1–9. 10.1016/j.ibiod.2017.05.020

Mallick, N. (2002). Biotechnological potential of immobilized algae for wastewater N, P and metal removal: A review. In: *BioMetals*, *15*.

Mantzavinos, D., & Psillakis, E. (2004). Enhancement of biodegradability of industrial wastewaters by chemical oxidation pre-treatment. *Journal of Chemical Technology & Biotechnology*, *79*(5), 431–454. 10.1002/jctb.1020

Markou, G., Angelidaki, I., & Georgakakis, D. (2012). Microalgal carbohydrates: An overview of the factors influencing carbohydrates production, and of main bio-conversion technologies for production of biofuels. *Applied Microbiology and Biotechnology*, *96*(3), 631–645. 10.1007/s00253-012-4398-0

Martínez, M. E., Sánchez, S., Jiménez, J. M., el Yousfi, F., & Muñoz, L. (2000). Nitrogen and phosphorus removal from urban wastewater by the microalga Scenedesmus ob-liquus. *Bioresource Technology*, *73*(3), 263–272. 10.1016/S0960-8524(99)00121-2

Masseret E., Ambrald C., Bourdier G., & Sargos D. (2000). Effects of a waste stabili-zation lagoon discharge on bacterial and phytoplanktonic communities of a stream. *Water Environment Research*, 72, 285–294.

Mata, T. M., Martins, A. A., & Caetano, N. S. (2010). Microalgae for biodiesel pro-duction and other applications: A review. *Renewable and Sustainable Energy Reviews*, 14(1), 217–232. Pergamon. 10.1016/j.rser.2009.07.020

Mccormick, K., & Kautto, N. (2013). The bioeconomy in Europe: An overview. *Sustainability*, 5, 2589–2608. 10.3390/su5062589

McKendry, P. (2002). Energy production from biomass (part 1): Overview of biomass. *Bioresource Technology*, *83*(1), 37–46. 10.1016/S0960-8524(01)00118-3

Meng, Y., Luan, F., Yuan, H., Chen, X., & Li, X. (2017). Enhancing anaerobic digestion performance of crude lipid in food waste by enzymatic pretreatment. *Bioresource Technology*, *224*, 48–55. 10.1016/j.biortech.2016.10.052

Moqsud, M. A., Omine, K., Yasufuku, N., Hyodo, M., & Nakata, Y. (2013). Microbial fuel cell (MFC) for bioelectricity generation from organic wastes. *Waste Management*, *33*(11), 2465–2469. 10.1016/j.wasman.2013.07.026

Moqsud, M. A., Yoshitake, J., Bushra, Q. S., Hyodo, M., Omine, K., & Strik, D. (2015). Compost in plant microbial fuel cell for bioelectricity generation. *Waste Management*, *36*, 63–69. 10.1016/j.wasman.2014.11.004

Mulbry, W., Kangas, P., & Kondrad, S. (2010). Toward scrubbing the bay: Nutrient removal using small algal turf scrubbers on Chesapeake Bay tributaries. *Ecological Engineering*, *36*(4), 536–541. 10.1016/j.ecoleng.2009.11.026

Mulbry, W., Kondrad, S., & Buyer, J. (2008). Treatment of dairy and swine manure effluents using freshwater algae: Fatty acid content and composition of algal biomass at different manure loading rates. *Journal of Applied Phycology*, *20*, 1079–1085. 10.1007/s10811-008-9314-8

Mulbry, W. W., & Wilkie, A. C. (2001). Growth of benthic freshwater algae on dairy manures. *Journal of Applied Phycology*, *13*, 301–306.

Muñoz, R., & Guieysse, B. (2006). Algal-bacterial processes for the treatment of hazardous contaminants: A review. *Water Research*, *40*(15), 2799–2815. Elsevier Ltd. 10.1016/j.watres.2006.06.011

Muñoz, R., Köllner, C., & Guieysse, B. (2009). Biofilm photobioreactors for the treatment of industrial wastewaters. *Journal of Hazardous Materials*, *161*(1), 29–34. 10.1016/j.jhazmat.2008.03.018

Nastro, R. A. (2014). Microbial fuel cells in waste treatment: Recent advances. *International Journal of Performability Engineering*, *10*, 367. https://doi.org/10.23940/IJPE.14.4.P367.MAG

Nastro, R. A., Suglia, A., Pasquale, V., Toscanesi, M., Trifuoggi, M., Guida, M. (2014). Efficiency measures of polycyclic aromatic hydrocarbons bioremediation process through ecotoxicological tests. *International Journal of Performability Engineering*, *10*, 411–418.

Ochando-Pulido, J. M., Hodaifa, G., & Martinez-Ferez, A. (2012). Fouling inhibition upon fenton-like oxidation pretreatment for olive mill wastewater reclamation by membrane process. *Chemical Engineering and Processing: Process Intensification*, *62*, 89–98. 10.1016/j.cep.2012.09.004

Paddock, M. B. (2019). Microalgae wastewater treatment: A brief history. *Preprints*, *2019*, 120377. doi: 10.20944/preprints201912.0377.v1.

Pan, J. R., Huang, C., Jiang, W., & Chen, C. (2005). Treatment of wastewater containing nano-scale silica particles by dead-end microfiltration: Evaluation of pretreatment methods. *Desalination*, *179*(1–3, special issue), 31–40. 10.1016/j.desal.2004.11.053

Pandey, V. K., Anjum, N., & Chandra, R. (2016). Algae as a biofuel: Renewable source for liquid fuel. *Carbon – Science and Technology*, *8*(3), 86–93. http://www.applied-science-innovations.com

Park, J. B. K., Craggs, R. J., & Shilton, A. N. (2011). Wastewater treatment high rate algal ponds for biofuel production. *Bioresource Technology*, *102*(1), 35–42. 10.1016/j.biortech.2010.06.158

Parker, M. S., Mock, T., & Armbrust, E. V. (2008). Genomic insights into marine microalgae. *Annual Review of Genetics*, *42*, 619–645. 10.1146/annurev.genet.42.110807.091417

Perey, J. R., Chiu, P. C., Huang, C.-P., & Cha, D. K. (2002). Zero-valent iron pretreatment for enhancing the biodegradability of azo dyes. *Water Environment Research*, *74*(3), 221–225. 10.2175/106143002x139938

Pittman, J. K., Dean, A. P., & Osundeko, O. (2011). The potential of sustainable algal biofuel production using wastewater resources. *Bioresource Technology*, *102*(1), 17–25. 10.1016/j.biortech.2010.06.035

Ponce-Robles, L., Miralles-Cuevas, S., Oller, I., Agüera, A., Trinidad-Lozano, M. J., Yuste, F. J., & Malato, S. (2017). Cork boiling wastewater treatment and reuse through combination of advanced oxidation technologies. *Environmental Science and Pollution Research*, *24*(7), 6317–6328. 10.1007/s11356-016-7274-0

Powell, E. E., Evitts, R. W., Hill, G. A., & Bolster, J. C. (2011). A microbial fuel cell with a photosynthetic microalgae cathodic half cell coupled to a yeast anodic half cell. *Energy Sources, Part A: Recovery, Utilization and Environmental Effects*, *33*(5), 440–448. 10.1080/15567030903096931

Powell, N., Shilton, A., Pratt, S., & Chisti, Y. (2011). Luxury uptake of phosphorus by microalgae in full-scale waste stabilisation ponds. *Water Science and Technology*, *63*(4), 704–709. 10.2166/wst.2011.116

Prabakar, D., Suvetha K. S., Manimudi, V. T., Mathimani, T., Kumar, G., Rene, E. R., & Pugazhendhi, A. (2018). Pretreatment technologies for industrial effluents: Critical review on bioenergy production and environmental concerns. *Journal of Environmental Management*, *218*, 165–180. 10.1016/j.jenvman.2018.03.136

Prathima Devi, M., Venkata Subhash, G., & Venkata Mohan, S. (2012). Heterotrophic cultivation of mixed microalgae for lipid accumulation and wastewater treatment during sequential growth and starvation phases: Effect of nutrient supplementation. *Renewable Energy*, *43*, 276–283. 10.1016/j.renene.2011.11.021

Pugazhendhi, A., Nagappan, S., Bhosale, R. R., Tsai, P. C., Natarajan, S., Devendran, S., Al-Haj, L., Ponnusamy, V. K., & Kumar, G. (2020). Various potential techniques to reduce the water footprint of microalgal biomass production for biofuel—A review. *Science of the Total Environment*, 749, 142218. 10.1016/j.scitotenv.2020.142218. Elsevier B.V.

Putt, R., Singh, M., Chinnasamy, S., & Das, K. C. (2011). An efficient system for carbonation of high-rate algae pond water to enhance CO_2 mass transfer. *Bioresource Technology*, *102*(3), 3240–3245. 10.1016/j.biortech.2010.11.029

Rahimnejad, M., Adhami, A., Darvari, S., Zirepour, A., & Oh, S. E. (2015). Microbial fuel cell as new technology for bioelectricity generation: A review. *Alexandria Engineering Journal*, 54 (3), 745–756. Elsevier B.V. 10.1016/j.aej.2015.03.031

Raj, T., Chandrasekhar, K., Banu, R., Yoon, J.-J., Kumar, G., Kim, S.-H. (2021a). Synthesis of γ-valerolactone (GVL) and their applications for lignocellulosic deconstruction for sustainable green biorefineries. *Fuel*, *303*, 121333. https://doi.org/10.1016/J.FUEL.2021.121333

Raj, T., Chandrasekhar, K., Kumar, A. N., & Kim, S.-H. (2021b). Recent biotechnological trends in lactic acid bacterial fermentation for food processing industries. *Systems Microbiology and Biomanufacturing*, 2, 1–27. https://doi.org/10.1007/S43393-021-00044-W

Rashid, N., Cui, Y. F., Muhammad, S. U. R., & Han, J. I. (2013). Enhanced electricity generation by using algae biomass and activated sludge in microbial fuel cell. *Science of the Total Environment*, *456–457*, 91–94. 10.1016/j.scitotenv.2013.03.067

Rittmann, B. E. (2008). Opportunities for renewable bioenergy using microorganisms. *Biotechnology and Bioengineering*, *100*(2), 203–212. 10.1002/bit.21875

Robson, P., Jensen, E., Hawkins, S., White, S. R., Kenobi, K., Clifton-Brown, J., Donnison, I., & Farrar, K. (2013). Accelerating the domestication of a bioenergy crop: Identifying and modelling morphological targets for sustainable yield increase in Miscanthus. *Journal of Experimental Botany, 64*(14), 4143–4155. 10.1093/jxb/ert225

Ruffino, B., Campo, G., Genon, G., Lorenzi, E., Novarino, D., Scibilia, G., & Zanetti, M. (2015). Improvement of anaerobic digestion of sewage sludge in a wastewater treatment plant by means of mechanical and thermal pre-treatments: Performance, energy and economical assessment. *Bioresource Technology, 175*, 298–308. 10.1016/j.biortech.2014.10.071

Ruiz-Marin, A., Mendoza-Espinosa, L. G., & Stephenson, T. (2010). Growth and nutrient removal in free and immobilized green algae in batch and semi-continuous cultures treating real wastewater. *Bioresource Technology, 101*(1), 58–64. 10.1016/j.biortech.2009.02.076

Saba, B., Christy, A. D., Yu, Z., & Co, A. C. (2017). Sustainable power generation from bacterio-algal microbial fuel cells (MFCs): An overview. *Renewable and Sustainable Energy Reviews, 73*, 75–84. 10.1016/j.rser.2017.01.115. Elsevier Ltd.

Saha, M., Eskicioglu, C., & Marin, J. (2011). Microwave, ultrasonic and chemo-mechanical pretreatments for enhancing methane potential of pulp mill wastewater treatment sludge. *Bioresource Technology, 102*(17), 7815–7826. 10.1016/j.biortech.2011.06.053

Saifullah, A. Z., Abdul Karim, M., & Ahmad-Yazid, A. (2014). Microalgae: An alternative source of renewable energy. *American Journal of Engineering Research, 03*, 330–338.

Sangave, P. C., & Pandit, A. B. (2006). Enhancement in biodegradability of distillery wastewater using enzymatic pretreatment. *Journal of Environmental Management, 78*(1), 77–85. 10.1016/j.jenvman.2005.03.012

Schenk, P. M., Thomas-Hall, S. R., Stephens, E., Marx, U. C., Mussgnug, J. H., Posten, C., Kruse, O., Hankamer, B., Schenk, P. M., Thomas-Hall, S. R., Stephens, E., Marx, U. C., Hankamer, B., Mussgnug, J. H., Kruse, O., & Posten, C. (2008). Second generation biofuels: High-efficiency microalgae for biodiesel production. *BioEnergy Research, 1*, 20–43. 10.1007/s12155-008-9008-8

Schnürer, A. (2016). Biogas production: Microbiology and technology. *Advances in Biochemical Engineering/Biotechnology, 156*, 195–234. 10.1007/10_2016_5. Springer Science and Business Media Deutschland GmbH.

Sharma, Y. C., Singh, B., & Korstad, J. (2011). A critical review on recent methods used for economically viable and eco-friendly development of microalgae as a potential feedstock for synthesis of biodiesel. Green Chemistry, *13*, 2993–3006. https://doi.org/10.1039/C1GC15535K

Sheehan, J., Dunahay, T., Benemann, J., & Roessler, P. (1978). *A look back at the U.S. Department of Energy's Aquatic Species Program: Biodiesel from Algae Close-Out Report.*

Shen, J., Ou, C., Zhou, Z., Chen, J., Fang, K., Sun, X., Li, J., Zhou, L., & Wang, L. (2013). Pretreatment of 2,4-dinitroanisole (DNAN) producing wastewater using a combined zero-valent iron (ZVI) reduction and Fenton oxidation process. *Journal of Hazardous Materials, 260*, 993–1000. 10.1016/j.jhazmat.2013.07.003

Shi, J., Podola, B., & Melkonian, M. (2007). Removal of nitrogen and phosphorus from wastewater using microalgae immobilized on twin layers: an experimental study. *Journal of Applied Phycology, 19*, 417–423. https://doi.org/10.1007/s10811-006-9148-1

Shikha, K., & Rita, C. Y. (2012). Biodiesel production from non edible-oils: A review. *Journal of Chemical and Pharmaceutical Research, 2012*(9), 4219–4230. www.jocpr.com

Singh, S. P., & Singh, P. (2014). Effect of CO_2 concentration on algal growth: A review. *Renewable and Sustainable Energy Reviews*, 38, 172–179. 10.1016/j.rser.2014.05 .043. Elsevier Ltd.

Sivagurunathan, P., Anburajan, P., Kumar, G., Arivalagan, P., Bakonyi, P., & Kim, S. H. (2017). Improvement of hydrogen fermentation of galactose by combined inoculation strategy. *Journal of Bioscience and Bioengineering*, 123(3), 353–357. 10.1016/j.jbiosc.2016.10.006

Slade, R., & Bauen, A. (2013). Micro-algae cultivation for biofuels: Cost, energy balance, environmental impacts and future prospects. *Biomass and Bioenergy*, 53, 29–38. 10.1016/j.biombioe.2012.12.019

Sri Bala Kameswari, K., Chitra Kalyanaraman, & Thanasekaran, K. (2011). Effect of ozonation and ultrasonication pretreatment processes on co-digestion of tannery solid wastes. *Clean Technologies and Environmental Policy*, 13(3), 517–525. 10.1007/s10098-010-0334-0

Stoller, M. (2009). On the effect of flocculation as pretreatment process and particle size distribution for membrane fouling reduction. *Desalination*, 240(1–3), 209–217. 10.1016/j.desal.2007.12.042

Sun, L., Liu, T., Müller, B., & Schnürer, A. (2016). The microbial community structure in industrial biogas plants influences the degradation rate of straw and cellulose in batch tests. *Biotechnology for Biofuels*, 9, 128. 10.1186/s13068-016-0543-9

Sun, Y., & Cheng, J. J. (2005). Dilute acid pretreatment of rye straw and bermudagrass for ethanol production. *Bioresource Technology*, 96(14), 1599–1606. 10.1016/ j.biortech.2004.12.022

Tan, C. H., Show, P. L., Chang, J. S., Ling, T. C., & Lan, J. C. W. (2015). Novel approaches of producing bioenergies from microalgae: A recent review. *Biotechnology Advances*, 33(6), 1219–1227. 10.1016/j.biotechadv.2015.02.013. Elsevier Inc.

Travieso, L., Benitez, F., & Dupeiron, R. (1992). Sewage treatment using immobilied microalgae. *Bioresource Technology*, 40(2), 183–187. 10.1016/0960-8524(92)902 07-E

Valladão, A. B. G., Freire, D. M. G., & Cammarota, M. C. (2007). Enzymatic pre-hydrolysis applied to the anaerobic treatment of effluents from poultry slaughterhouses. *International Biodeterioration and Biodegradation*, 60(4), 219–225. 10.1016/ j.ibiod.2007.03.005

van der Weijde, T., Alvim Kamei, C. L., Torres, A. F., Vermerris, W., Dolstra, O., Visser, R. G. F., & Trindade, L. M. (2013). The potential of C4 grasses for cellulosic biofuel production. *Frontiers in Plant Science*, 4(May), 107. https://doi.org/10.3389/ fpls.2013.00107. Frontiers Research Foundation.

Vasumathi, K. K., Premalatha, M., & Subramanian, P. (2012). Parameters influencing the design of photobioreactor for the growth of microalgae. *Renewable and Sustainable Energy Reviews*, 16(7), 5443–5450. https://doi.org/10.1016/j.rser.2012.06.013. Pergamon.

Veluchamy, C., & Kalamdhad, A. S. (2017). Enhanced methane production and its kinetics model of thermally pretreated lignocellulose waste material. *Bioresource Technology*, 241, 1–9. https://doi.org/10.1016/j.biortech.2017.05.068

Venkata Mohan, S., & Chandrasekhar, K. (2011a). Self-induced bio-potential and graphite electron accepting conditions enhances petroleum sludge degradation in bioelectrochemical system with simultaneous power generation. *Bioresource Technology*, 102, 9532–9541.

Venkata Mohan, S., & Chandrasekhar, K. (2011b). Solid phase microbial fuel cell (SMFC) for harnessing bioelectricity from composite food waste fermentation:

Influence of electrode assembly and buffering capacity. *Bioresource Technology*, *102*, 7077–7085.

Venkata Mohan, S., Chandrasekhar, K., Chiranjeevi, P., & Babu, P. S. (2013). Chapter 10 – Biohydrogen production from wastewater A2 – Pandey, Ashok. In: J.-S. Chang, P. C. Hallenbecka, C. Larroche (Eds.), *Biohydrogen* (pp. 223–257). Elsevier, Amsterdam. https://doi.org/10.1016/B978-0-444-59555-3.00010-6

Venkata Mohan, S., Prathima Devi, M., Venkateswar Reddy, M., Chandrasekhar, K., Asha Juwarkar, & Sarma, P. N. (2011). Bioremediation of real field petroleum sludge by mixed consortia under anaerobic conditions: Influence of biostimulation and bioaugmentation. *Environmental Engineering and Management Journal*, *10*(11), 1609–1616.

Venkata Mohan, S., Srikanth, S., Chiranjeevi, P., Arora, S., & Chandra, R. (2014). Algal biocathode for in situ terminal electron acceptor (TEA) production: Synergetic association of bacteria-microalgae metabolism for the functioning of biofuel cell. *Bioresource Technology*, *166*, 566–574. https://doi.org/10.1016/j.biortech.2014.05.081

Venkata Mohan, S., Vijaya Bhaskar, Y., Murali Krishna, P., Chandrasekhara Rao, N., Lalit Babu, V., & Sarma, P. N. (2007). Biohydrogen production from chemical wastewater as substrate by selectively enriched anaerobic mixed consortia: Influence of fermentation pH and substrate composition. *International Journal of Hydrogen Energy*, *32*(13), 2286–2295. https://doi.org/10.1016/j.ijhydene.2007.03.015

Viswanath, B., & Bux, F. (2012). Biodiesel production potential of wastewater microalgae chlorella sp. under photoautotrophic and heterotrophic growth conditions. *British Journal of Engineering and Technology*, *1*(1), 251–264.

Voltolina, D., Cordero, B., Nieves, M., & Soto, L. P. (1999). Growth of *Scenedesmus* sp. in artificial wastewater. *Bioresource Technology*, *68*(3), 265–268. https://doi.org/1 0.1016/S0960-8524(98)00150-3

Wang, L., Li, Y., Chen, P., Min, M., Chen, Y., Zhu, J., & Ruan, R. R. (2010). Anaerobic digested dairy manure as a nutrient supplement for cultivation of oil-rich green microalgae Chlorella sp. *Bioresource Technology*, *101*(8), 2623–2628. https://doi.org/10.1016/j.biortech.2009.10.062

Wang, S., Hou, X., & Su, H.(2017). Exploration of the relationship between biogas production and microbial community under high salinity conditions. *Scientific Reports*, 7, https://doi.org/10.1038/s41598-017-01298-y

Wehr, J. D. (2007). Algae: Anatomy, biochemistry, and biotechnology by Barsanti, L. & Gualtieri, P. *Journal of Phycology*, *43*(2), 412–414. https://doi.org/10.1111/j.1529-8817.2007.00335.x

Wen, X., Du, K., Wang, Z., Peng, X., Luo, L., Tao, H., Xu, Y., Zhang, D., Geng, Y., & Li, Y. (2016). Effective cultivation of microalgae for biofuel production: A pilot-scale evaluation of a novel oleaginous microalga Graesiella sp. WBG-1. *Biotechnology for Biofuels*, *9*(1), 123. https://doi.org/10.1186/s13068-016-0541-y

Wilkie, A. C., & Mulbry, W. W. (2002). Recovery of dairy manure nutrients by benthic freshwater algae. *Bioresource Technology*, *84*(1), 81–91. https://doi.org/10.1016/S0960-8524(02)00003-2

Wirth, R., Kovács, E., Maróti, G., Bagi, Z., Rákhely, G., & Kovács, K. L. (2012). Characterization of a biogas-producing microbial community by short-read next generation DNA sequencing. *Biotechnology for Biofuels*, 5, 41. https://doi.org/10.1186/1754-6834-5-41

Wrigley, T. J., & Toerien, D. F. (1990). Limnological aspects of small sewage ponds. *Water Research*, *24*(1), 83–90. https://doi.org/10.1016/0043-1354(90)90068-H

Xin, L., Hong-Ying, H., & Jia, Y. (2010). Lipid accumulation and nutrient removal properties of a newly isolated freshwater microalga, Scenedesmus sp. LX1, growing in secondary effluent. *New Biotechnology*, *27*(1), 59–63. https://doi.org/10.1016/j.nbt.2009.11.006

Xiong, Z., Cao, J., Yang, D., Lai, B., & Yang, P. (2017). Coagulation-flocculation as pretreatment for micro-scale Fe/Cu/O$_3$ process (CF-mFe/Cu/O$_3$) treatment of the coating wastewater from automobile manufacturing. *Chemosphere*, *166*, 343–351. https://doi.org/10.1016/j.chemosphere.2016.09.038

Zhang, E., Wang, B., Wang, Q., Zhang, S., & Zhao, B. (2008). Ammonia-nitrogen and orthophosphate removal by immobilized Scenedesmus sp. isolated from municipal wastewater for potential use in tertiary treatment. *Bioresource Technology*, *99*(9), 3787–3793. https://doi.org/10.1016/j.biortech.2007.07.011

Zhang, J., Li, W., Lee, J., Loh, K. C., Dai, Y., & Tong, Y. W. (2017). Enhancement of biogas production in anaerobic co-digestion of food waste and waste activated sludge by biological co-pretreatment. *Energy*, *137*, 479–486. https://doi.org/10.1016/j.energy.2017.02.163

Zhen-Feng, S., Xin, L., Hong-Ying, H., Yin-Hu, W., & Tsutomu, N. (2011). Culture of Scenedesmus sp. LX1 in the modified effluent of a wastewater treatment plant of an electric factory by photo-membrane bioreactor. *Bioresource Technology*, *102*(17), 7627–7632. https://doi.org/10.1016/j.biortech.2011.05.009

Zhou, W., Li, Y., Min, M., Hu, B., Chen, P., & Ruan, R. (2011). Local bioprospecting for high-lipid producing microalgal strains to be grown on concentrated municipal wastewater for biofuel production. *Bioresource Technology*, *102*(13), 6909–6919. https://doi.org/10.1016/j.biortech.2011.04.038

Zych, D. (2008). *The viability of corn cobs as a bioenergy feedstock.*

3 Biofilm and the Electron Transfer Mechanism in Bioelectrochemical Systems

Hany Abd El-Raheem
Zewail City of Science and Technology, Giza, Egypt

Université Paris-Saclay, CNRS, Institut de Chimie
Moléculaire et des Matériaux d'Orsay (ICMMO), ECBB,
Orsay, France

Hafsa Korri-Youssoufi
Université Paris-Saclay, CNRS, Institut de Chimie
Moléculaire et des Matériaux d'Orsay (ICMMO), ECBB,
Orsay, France

Rabeay Y.A. Hassan
Zewail City of Science and Technology, Giza, Egypt

Applied Organic Chemistry Department, National
Research Centre (NRC), Dokki, Giza, Egypt

CONTENTS

DOI: 10.1201/9781003225430-3

3.1 INTRODUCTION TO BIOFILM

Biofilms, as an expression, refer to a biofilm-encased community of microorganisms (bacteria, algae, fungi, or viruses), and can be described as a three-dimensional complex community of microbial cells that are irreversibly associated with inert or living solid surfaces and embedded in an exopolysaccharide (EPS) matrix formed of extracellular polymeric substances which are considered as a cross-linked polymer gel (Azeredo et al., 2017a; Barsoumian et al., 2015a; Donlan & Costerton, 2002; Ritenberg et al., 2016; Sultana et al., 2015). The EPS provides a safeguarding barrier to the microbial communities from antibiotics and other toxic substances (Donlan & Costerton, 2002; Minaev, 2007). The EPS consists mainly of organic heterogeneous macromolecules including lipids, proteins, nucleic acids, and polysaccharides separated by water channels (Flemming & Wingender, 2010b; Whitchurch et al., 2002). Water channels are essential for the persistence of biofilm formed, allowing the movement of nutrients and metabolic products to reach the biofilm community (Donlan & Costerton, 2002; Sousa et al., 2011).

Biofilms are sensitive to temperature, pH, salinity, humidity, nutrients, and osmolarity (Kaali et al., 2011; Magot et al., 2000). They can exist in human tissues such as heart valves, vaginal surfaces, and teeth (Hall-Stoodley et al., 2004). Biofilms are useful in electrical energy generation as in microbial fuel cells (MFCs) (Mahmoud et al., 2018; Mahmoud et al., 2021; Sedki et al., 2019), bioremediation processes, and wastewater treatment procedures (Singh et al., 2006; Mattila-Sandholm & Wirtanen, 1992).

The discovery of biofilms was in 1683 when the Dutch microbiologist Antonie van Leeuwenhoek observed and described microbial aggregates on his own mouth by using his primitive microscope. He saw aggregated biofilm in the "scurf of the teeth" and from "particles scraped off his tongue". In 1933, Henrici studied biofouling in freshwater by using direct microscopy and observed that most parts of water bacteria are not free-floating organisms, but grow attached upon submerged surfaces. This is the first recorded evidence of a biofilm's existence (Henrici, 1933). ZoBell and Allen studied the growth and adherence of bacteria on submerged glass slides in seawater (Zobell & Allen, 1935). In 1981, dentists from the University of Lund, Sweden, published the first two medical reports using the term of "biofilm" (Jendresen & Glantz, 1981; Jendresen et al., 1981). Subsequently, J. W. Costerton introduced the term "biofilm" growth to the field of medical microbiology in 1985 (Costerton et al., 1974; Nickel et al., 1985). Due to the global concern, the first conference about biofilm was organized by J. W. Costerton in Snowbird, Utah, USA, in 1996.

3.2 BACTERIAL BIOFILMS

Bacterial communities exist in two various forms, planktonic state (free-floating) or/and sessile state (adhered to a surface) (Hall-Stoodley & Stoodley, 2009; Stoodley et al., 2002). Bacterial biofilms are involved in various chronic infections in human tissues and indwelling medical devices in particular (Sabir et al., 2017; Davies & Marques, 2009). In the biofilm matrix, bacterial behaviors are coordinated by cell-cell communication using secreted small molecules (quorum sensing molecules) that allow bacteria to sense and respond to the environment. Biofilms are more resistant to environmental stresses, such as metals, toxicity, dehydration, and UV light exposure than planktonic cells. The biofilms are a mixture of heterogeneous communities of microbial cells surrounded by a condensed layer of exopolysaccharides matrix (EPSs), and are strongly adhered to living tissues or solid surfaces (Flemming & Wingender, 2010a; Saratale et al., 2017a). In most biofilms, the living cells represent less than 10% of the total content in prevalence represented by the matrix (about 90%). In general, the biofilm matrices contain components such as polysaccharides, proteins, and extracellular DNA, but their content is dependent on the bacterial species and the environmental conditions (Azeredo et al., 2017b). Moreover, it provides high mechanical stability, the EPS environment mediates the cell-cell adhesion to the solid surfaces, and forms a cohesive three-dimensional network that interconnects biofilm cells (Pandit et al., 2018a). Thus, the formation of biofilms is one of the major problems around environmental and biological surfaces and causes different challenges in various biomedical science fields (Butler & Boltz, 2014; Chaturongkasumrit et al., 2011; Srey et al., 2013). For example, microbial contamination on metal implants and prosthetic biomedical devices causing biofilm formation can be life-threatening, leading to chronic infections, device failure, and high mortality rates (Magana et al., 2018; Veerachamy et al., 2014a).

3.3 DESCRIPTION OF THE STEPS OF BIOFILM FORMATION

The formation of biofilm involves various stages of a process. They are the initial attachment, microcolony formation, maturation and architecture, and dispersal of biofilm from the surface (Büttner et al., 2015; Gu et al., 2013; Joo & Otto, 2012; Klausen et al., 2003; LewisOscar et al., 2016). Once the bacteria adhere to the surface, they can initiate this series of reactions that would result in biofilm formation. The different stages of biofilm formation are ordered in the following steps (Figure 3.1). Thus, biofilm formation is commonly considered to occur in four main phases: (1) microbial initial attachment to a surface, (2) microcolony formation, (3) biofilm development, and (4) dispersal (detachment) of living microbes to colonize new areas. Depending on the cell adherence mechanism, the first step of "initial attachment" in the process can be either active or passive adhesion (Dufour et al., 2010; O'Toole et al., 2000; Srey et al., 2013). The microbial cells start to attach with the surface through their adhesive surface structures called appendages like Flagella and *Pilli* and promote the "active adhesion". This active attachment provides a binding force between the microbial cells and the surface of attachment (Berne et al., 2015; Chmielewski & Frank, 2003). Flagella and pili play a role in surface recognizing by which microorganisms sense and respond to contact with the surface (O'Toole & Wong, 2016). In contrast, the "passive adhesion" is enabled by gravity, diffusion, and fluid dynamics (Chmielewski & Frank, 2003). Initial attachment (the first phase of biofilm formation) is a reversible stage due to weak interactions between the bacteria and surface; hence, at this stage, living cells are still able to detach and return to the planktonic shape (Büttner

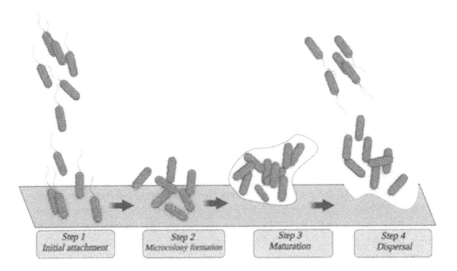

| Step 1 | Step 2 | Step 3 | Step 4 |
| Initial attachment | Microcolony formation | Maturation | Dispersal |

FIGURE 3.1 Stages of biofilm formation on solid surfaces such as metallic stents, catheters, or any other implantable devices.

et al., 2015; Floyd et al., 2017; Veerachamy et al., 2014b). Surface charges show a major contribution in cell surface interaction (Song et al., 2015). Moreover, the nature of the surfaces affects the attachment, a smooth surface provides a lower adhesion, while a rough surface provides a higher adhesion (Garrett et al., 2008). Hydrophobic surfaces like plastics and Teflon have a significant role in strengthening the attachment of microbes, this is due to reducing the force of repulsion between the surface and the bacteria (Kumar & Anand, 1998). Post the attachment to surfaces, microbial cells start a new phase of multiplication, initiated through a particular chemical signaling within the EPS, which leads to the formation of micro-colonies. At this step, microorganisms display a coordinated behavior through cell-cell communication commonly known as quorum sensing (QS), which is important for monitoring virulence and biofilm formation along with the sporulation, factor secretion, competence. Accordingly, the adhered microbial cells grow and mature by communicating with one another through the production of auto-inducer signals which consequences in the expression of biofilm-specific genes (Gupta et al., 2016). These auto-inducer signals facilitate quorum sensing molecule production (Federle & Bassler, 2003). In this stage, cells start the secretion of the EPS that surrounds the cells (Parsek & Singh, 2003). During their aggregation and accumulation, several layers of cell clusters are formed on the surface that gets enclosed within the EPS where inter-cellular signaling and quorum sensing takes place (Gupta et al., 2016; Veerachamy et al., 2014b). Thus, maturation is conducted by two steps (I) – involving inter-cell communication and the production of auto-inducer signals and (II) – increasing the micro-colony extent and thickness to value around 100 mm (Paharik & Horswill, 2016; Wei et al., 2015).

In summary, the maturation stage involves EPS production, aggregation of cells, chemical interactions, quorum sensing, and formation of micro- and macro-colonies (Von Eiff et al., 2005). The last stage of biofilm formation is cell dispersal or detachment. This process is essential for the spread of individual cells or cell clusters into new locations, from sessile into motile form (Kaplan, 2010). It is believed to occur in response to several mechanisms that promote cellular dispersion, such as enzyme digestion of the extracellular matrix, sheer mechanical forces, accumulation of metabolites, the depletion of oxygen, as well as some external forces (McDougald et al., 2012). However, some types of bacteria do not produce extracellular polysaccharides and the bacterial cells disperse directly into the outer environment (Baselga et al., 1994). Detachment of microbial cells and movement to a new location support the spread of chronic infections and other severities (Otto, 2013; Veerachamy et al., 2014b). This process is often referred to as metastatic seeding (Bjarnsholt, 2013; Chao et al., 2015; Von Eiff et al., 2005). Interestingly, blocking the respiratory chain or the electron transfer chain of the living microorganisms leads to the prevention of biofilm formation, as has been studied on the *Candida albicans* (Thibane et al., 2010). In terms of that, reactive oxygen species (ROS) that arise from blocking the classical respiratory pathways have harmful and damaging effects on the matrix of biofilms (Liang et al., 2010). Eventually, a better understanding of microbial biofilm formation and its regulation is necessary to manage and/or to eliminate biofilm-related infections and of course will assist in providing effective anti-biofilm agents.

3.4 METHODS OF BIOFILM ASSESSMENT

The complex architecture dynamics of biofilms create challenges for routine measurements concerning the biofilm morphology, mass accumulation, number of viable cells, and other critical features. The assessments themselves are not the challenge; the lack of standardized protocols for evaluation is a big problem. Assessments can include the determination of the total number of cells (dead vs. alive), total organic carbon, or the total dry mass of the whole biofilm matrix. Biofilm morphology analysis can involve two-dimensional surface structures illuminated over staining techniques and light microscopy or three-dimensional topographies revealed by confocal scanning laser microscopy (CSLM). A suitable choice of techniques is built on the information needed, availability of equipment, and the cost of reagents. In the following sections, we will review the most commonly used methods of biofilm identification and characterization, which we categorize into direct methods and indirect methods.

3.4.1 DIRECT DETECTION METHODS

Biofilm can be investigated and assessed by direct imaging techniques such as light microscope (LM), transmission electron microscope (TEM), scanning electron microscope (SEM), confocal laser scanning microscope (CLSM), and the fluorescent microscope. These imaging techniques are used to visualize 3D structures and check the formation and existence of biofilms (Roy et al., 2018).

3.4.1.1 Light Microscopy (LM)

Microscopic techniques are the most used for the study of microorganisms. Among the oldest of these techniques is light microscopy (Haguenau et al., 2003). LM is the simplest, easiest, convenient, and fastest method to observe the formation and semi-quantitatively estimate the amount of biofilm formation adhered to surfaces (Azeredo et al., 2017a). The characterization of biofilm includes morphology (colonized or planktonic forms), abundance, size, and motility that can be provided by LM. However, 3D characterization is not possible using LM. Specific labels dyes can be used to provide fluorescent and epifluorescence which enhance the visualization of the recorded images (Christensen et al., 2000).

3.4.1.2 Transmission Electron Microscope (TEM)

Transmission electron microscopy (TEM) has been used for the first time to study the naturally occurring microbial films in aquatic systems and employed to characterize slime films growing in contaminated streams. Basically, the TEM affords imaging with a very higher resolution of up to 0.2 nm for the surface analysis of microbial cells and the surrounding environment. The satisfactory resolution for most reported biofilm studies ranged between 2 to 20 nm (Egerton, 2005). TEM performs a thin section of biofilm, so types of microorganisms present in biofilms, extracellular polymeric substance, and conditioning film may be detected.

3.4.1.3   Scanning Electron Microscope (SEM)

Scanning electron microscopy (SEM) is one of the common electron micro-scopic techniques exploited in biofilm 3D imaging since 1980s. The SEM was first served to characterize the growth of *Sphaerotilus natans* and its adherence to catheter lumen in a continuous flow recycle system. SEM can effectively vi-sualize the surface of the biofilm with very high magnification and excellent resolution. In that way, a complete shape of the organisms composing the biofilm in addition to their arrangement to each other and their connection to the ex-tracellular matrix can be easily analyzed. Before electron illumination, the outer layer of biofilm is sputtered with a very thin film of gold to facilitate visuali-zation of the biological matrix (Fassel & Edmiston, 2000).

3.4.1.4   Confocal Laser Scanning Microscopy (CLSM)

With the evolution of molecular techniques for the study of microbial com-munities in the 1980s and 1990s, confocal laser screening microscopy (CLSM) was rapidly developed and considered as a tool to visualize the biofilm three-dimensional (3D) morphology and physiology (Franklin et al., 2015; Shunmugaperumal, 2010). The CLSM can easily identify both biofilms components and localized microorganisms in the depth of thick samples. To facilitate visualization of the biofilm with CLSM, certain protocols were optimized for the biofilm formation using fluorescent protein (e.g. green fluor-escent protein (GFP)), which is a fluorescent protein expressed by localized mi-crobes within the biofilm matrix microorganisms (Shunmugaperumal, 2010).

3.4.1.5   Fluorescent in situ Hybridization (FISH)

The fluorescent in situ hybridization (FISH) technique has a powerful impact to quickly identify the heterogeneous complex structure biofilm community. Thus, phylogenetic identification could be conducted without the need to amplify their genes or to do further cultivation. On the other hand, The FISH affords rapid quantitative information about the abundance of microbial groups without PCR. FISH is relied on the identification of microorganisms using short (15 to 20 nucleotides) rRNA-complementary fluorescently labeled oligonucleotide probes (species, genes, or group-specific) that penetrate microbial cells, bind to RNA, and emit illuminated UV light (Kempf et al., 2000). The probe must be designed to label the conserved region of only a single species.

3.4.2   Indirect Detection Methods

The direct method determination of biofilm could be obtained by colorimetric method using various assay that involves various dyes that could attach the biofilm and gives a visible color observed by naked eyes. The ones most used are tube method, microtiter plate assay, and Congo red agar. This method gives an indication of the presence of biofilm; however, to determine the nature of the biofilm formed advanced molecular techniques should be used such as polymerase chain reaction (PCR).

3.4.2.1 Tube Method (TM)

Among the common biofilm producers, TM was described by Christensen et al. (1985) for the reliable qualitative detection of biofilm-producing microorganisms (Christensen et al., 1985). TM is a qualitative detection method it consists of observing biofilm lined on the bottom and walls of the tube marked with a dye. How does the TM assay take place? Microbial or clinical isolates are incubated for 24 h at 37°C in a polystyrene test tube that contained trypticase soy broth (TSB) as culture media. Planktonic cells are then removed or discharged from the tube by washing thoroughly with phosphate-buffered saline (PBS). Then the biofilms produced on the walls of the test tube are labeled for 1 h with the dye Safranin. Then, the dye-stained polystyrene test tube is rinsed twice with PBS to release excess or remained stain and let to dry. Eventually, the occurrence of visible film lined the walls on the bottom of the tube designates biofilm construction. The quantity of biofilm formation was elucidated according to the results of the control strain and graded visually as absent, moderate, and strong biofilm formation, respectively.

3.4.2.2 Microtiter Plate Assay (MPA)

The Microtiter plate method is a qualitative assay able to detect biofilm formation. It allows the observation of bacterial adherence to the surface. In this method, culture was incubated for 24 hours at 37°C. Biofilms made of microorganisms were attached on a plate bonded with sodium acetate which was stained by crystal violet dye, allowing the visualization of the biofilm (Stepanović et al., 2007). The advantages of MPA method are the simplicity, adaptability to small or large numbers of samples, the use of basic lab materials, and the variety of samples that can be tested in a single assay. The microtiter plate assay is predominantly useful for examining the early stages in biofilm formation, such as initial surface attachment (Redelman et al., 2012). The use of MPA is not limited to a certain protocol or using a specific dye for spectrophotometric determination. But, it can be considered a routine lab tool.

3.4.2.3 Congo Red Agar (CRA)

Qualitative assessment of biofilm could be conducted using the CRA that is which was described by Freeman in 1989 (Freeman et al., 1989). In this regard, a color change of colonies inoculated on a CRA medium is obtained and proportioned to the cell dentistry exited in the biofilm matrix. A mixture of Congo red dye (0.8 g/l), sucrose (36 g/l), and brain heart infusion (37 g/l) are the main components of the CRA. Under anaerobic conditions and incubation at 37°C for

24 hours, the morphological characterization of stained colonies that have undergone different colors are differentiated as biofilm producers.

3.4.2.4 Polymerase Chain Reaction (PCR)

The polymerase chain reaction (PCR) is a method widely used to study the biofilm, allowing the detection of genes associated with biofilm and individuate microbial colonization. Besides the very high specificity and sensitivity of PCR, the automatization is advantageous when compared to conventional culture-based detection methods. Several basic steps, such as sample preparation, enrichment media, selection of PCR primers, and adjustment of reaction time and temperature, must be conducted to obtain reliable PCR results.

3.4.2.5 Electrochemical Methods

The various mentioned methods allow the characterization of the various properties of biofilm-related to nature the morphology, molecular structure. The electrical behavior of biofilm is also needed to understand the electron exchange properties and the mechanism of electrical communication between the various components in the biofilm and the surrounding environment. Bio-electrochemistry is a powerful method that leads to explain and follow the electron transfer mechanism in biofilm. Thus, electrochemical characterization techniques such as voltammetry, potentiometry, Amperometry, and electrochemical impedance spectroscopy (EIS) are an alternative to microscopy, culture-based, and molecular techniques for detection of microbial biofilm (Xu et al., 2020). Electrochemical techniques have been defined as simple, easy to use, portable, cost effective, and disposable; all of these are features that make them ideal for point-of-care devices (Karunakaran et al., 2015). Electrochemical sensing is made possible by a typical three-electrode electrochemical cell consists of a working, a counter (CE), and a reference electrode. In these cases, the working electrode serves as a surface on which the biofilm formation takes place. Electrochemical techniques can characterize surface modifications upon biofilm growth by evaluating the electroactive area, the presence of electroactive microbial strains, or evaluating the rate of electrons exchange between biofilm and electrode. Cyclic voltammetry (CV) is the most common, simple, fast technique for acquiring qualitative and quantitative information on biological and microbial redox reactions (Hassan et al., 2017; Sedki et al., 2019).

The understanding of such properties could be exploited in various applications and devices such as biofuel cells or biosensors.

3.5 BIOFILM AND ELECTRON TRANSFERABILITY

3.5.1 DIRECT AND MEDIATED ELECTRON TRANSFER

Bioelectrochemical systems (BESs) are promising fast-expanding technologies that exploit microbial living systems, materials sciences, and electrochemistry to advance energy, environmental, and sensing strategies (Selim et al., 2017).

Accordingly, the BESs use the powerful catalytic activity of metabolically active cells to collect electrons from the biodegradable organic compounds that exist in their outer medium (Chandrasekhar & Venkata Mohan, 2012; Kumar et al., 2012; Mohan & Chandrasekhar, 2011a; Mohan & Chandrasekhar, 2011b). Hence, interaction(s) between living cells (electron donor) and solid conductive surfaces (anodes as electron acceptors) are implemented (Khater et al., 2015). A critical requirement for constructing a bioelectrochemical system is to sustain and enable direct electron transfer (DTE) between the afforded biofilm to the anode surface. Consequently, BESs can rely on: (i) extracellular electron transfer (i.e. microbial electrocatalytic performance providing extracellular electron transfer) or (ii) indirect communications through secreted metabolites, and quorum sensing molecules or any natural electrochemical active compounds produced by the inner-electrochemical microorganisms (Chandrasekhar et al., 2021c; Gambino et al., 2021; Hassan & Bilitewski, 2013).

The main pillars for BES constructions include the type of microorganisms used to construct the system, the bio-anode surface and its configuration, and the electrochemical reactor and its dimensions (Alfadaly et al., 2021; Hassan et al., 2021; Mahmoud et al., 2020b). In the BESs, the microbial electron transport chain (METC) is considered as the essential partition (Wesolowski et al., 2008). Therefore, the electron transfer from viable cells to anodes is exploited in microbial diagnoses for the rapid detection of pathogens (Cui et al., 2020; Hassan & Wollenberger, 2019; Mahmoud et al., 2020a), or, for bioelectricity generation using microbial fuel cells (MFCs) (Katuri et al., 2012; Ng et al., 2017). In terms of electron transfer, electroactive microbes can deliver their intracellular electrons directly (i.e. without any electron shuttles) to the electrode to produce electrochemical signals (Figure 3.2a). In 1911, for the first time, Michael C.

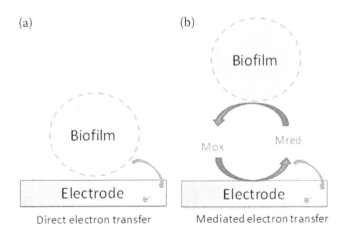

FIGURE 3.2 (a) Direct electron transfer from electrochemically active biofilm to the electrode surface; (b) mediated electron transfer from electrochemically inactive biofilm to electrode surface through providing artificial electron shuttles from redox mediators.

Potter demonstrated a current flow from a microbial culture, when he observed an electromotive force between electrodes immersed in a culture of bacteria or and electrodes immersed in a sterile environment. He claimed that the current was resulting from the breakdown of organic substances by the actions of living microbes (Potter, 1911). His finding was confirmed by the development of a stack of microbial fuel cells delivering 35 V (Cohen, 1931). On the other hand, electrochemically inactive biofilms could be integrated with electrode surfaces via extracellular electron receptors/transmitters or electron mediators (Figure 3.2b). Electron transmitters in the oxidized form can enter the living cell wall, as well as the cell membranes to capture the intracellular electrons. Subsequently, the reduced form of the electron shuttle is released back to deliver the accepted electrons through redox reactions occurring at the electrode surface (Chandrasekhar et al., 2015a; Chandrasekhar & Venkata Mohan, 2014a; Chandrasekhar & Venkata Mohan, 2014b). The use of soluble redox mediators in the BES is an operative to increase the rate of the extracellular electron transfer (Guo et al., 2020).

Metabolically active microbes can directly communicate with the anode over the formation of electroactive biofilms (Barsoumian et al., 2015b). In the microbial electrochemical systems, different biofilms could be characterized as electrochemically active or inactive (Halan et al., 2012). The electrochemically active biofilms are characterized as a microbial community of electrochemically active microorganisms interacting directly with conductive surfaces to transfer extracellular electrons (Babauta et al., 2012; Erable et al., 2010; Prévoteau & Rabaey, 2017). Therefore, physical connections through the microbial appendages, microbial nanowires, and natural cyt-c in certain microorganisms such as *Geobacter sulfurreducens, Shewanella oneidensis,* and *Thiobacillus denitrificans* are discovered (Reguera et al., 2006; Zhou et al., 2015).

Bioelectrochemical examination of the biofilm formation at modified electrode surfaces was conducted to understand the impact of the electrode constituents on the biofilm development (Cornejo et al., 2015; Mahmoud et al., 2020a) and to monitor a modification in the biofilm structures under different stresses through direct measurements of electron exchanges. The morphological characterization using microscopic techniques is generally associated to confirm the electrochemical process. For example, Sedki et al. demonstrated the effect of various nanostructured electrodes formed with reduced graphene oxide (rGO) and hyperbranched chitosan to assess the influence of the electrode materials on biofilm progression of *P. aeruginosa* along with monitoring its electrochemical/ morphological changes under different stresses (Figure 3.3). Its shows the *P. aeruginosa* cells at the carbon screen printed which no redox behavior due to the low connection between biofilm and electrode, (b) the slimy biofilm matrix of *P. aeruginosa* at the rGO-HBCs-CPE electrode after incubation for five days in LB media, and (c) a closer view showing the connection between biofilm bacterial cells through flagella and pili.

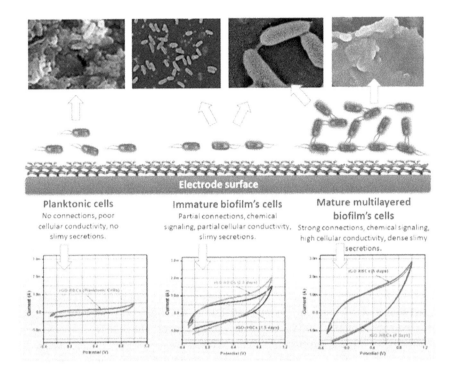

FIGURE 3.3   A schematic illustration of the biofilm growth was monitored by measuring the direct extracellular electron transferability of the biofilm matrix and the corresponding morphologies reproduced with permission from (Sedki et al., 2019).

3.5.2   Microbial Electron Exchanges Within a Microbial Community

Microbial community in a mixed culture can gain their bioenergy from exchangeable electron transfer with each other (Schink & Stams, 2006; Stams & Plugge, 2009). This kind of reaction is known as syntrophic, whereas an indirect mechanism is preferred, relying on hydrogen or format as electron transfer molecules (Figure 3.4). As an example, fermenting microorganisms produce hydrogen from organic materials, and archaea uptake the hydrogen for the production of biomethane. In some methanogenic communities, formate instead of hydrogen is acting as an electron transporter, while also syntrophic interactions with both formate and H_2-mediated IET have been reported (Dong & Stams, 1995).

3.5.3   Electron Transfer via Cable Bacteria

Sulfide bio-oxidation in the marine sediments (anoxic zone) is found to be coupled to the reduction of oxygen at the sediment surface. This kind of redox reaction is disconnected by a centimeter distance in the zone that depleted both compounds (Nielsen et al., 2010; Risgaard-Petersen et al., 2012). These

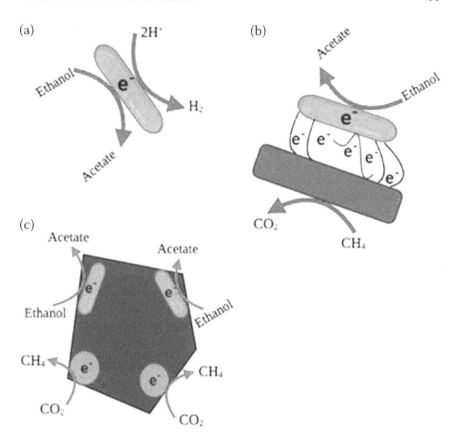

FIGURE 3.4 Three modes of electron transfer between microorganisms. (a) Via a soluble component such as formate or hydrogen; (b) direct contact; or (c) mediated electron transfer through a conductive abiotic carrier material.

discoveries directed the research attention to think about the existence of a generation of bio/or natural electric currents that could be involved in the oxidation-reduction reaction taking place in marine sediment (Nielsen et al., 2010).

To that end, cable bacteria (CB) are explored. CB is universally happening in multicellular filamentous bacteria that are electrochemically active or electrically conductive. The CB can transfer electrons from the oxidation of sulfide (on one-side) to reduce oxygen (on the other-side) over centimeter distances. Different than any other microorganisms known, The CB can divide the central energy-conserving redox reactions into two half reactions that happen in different cells as long as they are several centimeters apart. The evolutionary origin, molecular foundation, and genomic root of this exceptional metabolism were explored by Kasper U. Kjeldsen (Kjeldsen et al., 2019; Pfeffer et al., 2012).

3.6 APPLICATION OF BIOFILM IN BIOELECTROCHEMISTRY

3.6.1 MICROBIAL FUEL CELLS

A biofuel cell is a type of electrochemical cell, consisting of one or two chambers that are separated by an ion-exchange membrane or a diaphragm (Rabaey et al., 2007). The anode collects electrons from an oxidation reaction occurred by the attached microbes, while the cathode conveys electrons for the oxygen reduction reaction (ORR). Anode and cathode are typically made of conductive materials such as steel or carbon-based materials (Barrière & Downard, 2008; Guo et al., 2015; Rinaldi et al., 2008). Microbial fuel cells (MFCs) are common microbial electrochemical systems that exploit the activity of living-microorganisms to convert chemical energy through the oxidation of organic substrates to electricity (Chandrasekhar & Ahn, 2017; Chandrasekhar et al., 2015b; Deval et al., 2017; Dolch et al., 2014; Pandit et al., 2017; Saratale et al., 2017b). Typical MFCs consist of an anode and a cathode both of which are incubated with a liquid culture of living microbes (electrogenic or electroactive organisms). For the anodic reaction, living cells consume the degradable organic substrate (electron donors) that are transferred from the bulk solution and through the cellular metabolism to the anode surface (Chandrasekhar, 2019; Chandrasekhar et al., 2018; Kumar et al., 2018; Pandit et al., 2018b; Santoro et al., 2017; Venkata Mohan et al., 2019). Two different designs (i.e. single or double-chambers) of MFCs have been used, illustrated in Figure 3.5. The double chambers, an anaerobic anode chamber and an aerobic cathode chamber, are generally separated by a proton exchange membrane (PEM) such as sulfonated poly-(ether-ether-ketone) (SPEEK) as a kind of the non-fluorinated membranes or the Nafion as the most common PEM used (Shabani et al., 2020). Under limited oxygen conditions (hypoxic environment), the anodic inhalation is followed by biofilm formation at the solid electrode (Pasternak et al., 2018). A mixed/raw microbial culture or single microbial standard bacterial strain can be used to set up an MFC with high performance (Khater et al., 2018; Yi et al., 2018). Regulator factors, biofilm structure adaptation, and/or the type of the microbial community have a great impact on the electrochemical characteristics. The number and dimensions of the biofilm pores affect the transfer of metabolites to the electrode surface while the transfer of protons, substrate, and metabolites between the electrode surface and the bulk solution affect the current generation (Katuri et al., 2020). The current generated by MFCs arises from the transfer of electrons received by the anode when live bacteria oxidize organic materials, which are then transferred to the cathode via the external circuit (Chandrasekhar et al., 2021a; Chandrasekhar et al., 2021b), as shown in Figure 3.5. Typically, MFCs are classified into two types depending on how extracellular electrons are delivered from the attached microorganisms to the anode: (i) mediator-based MFCs, in which electro-active secreted metabolites or artificial redox compounds are used to shuttle the electrons (Li et al., 2020); (ii) mediated-less fuel cells do not require the addition of electroactive metabolites to

(a)

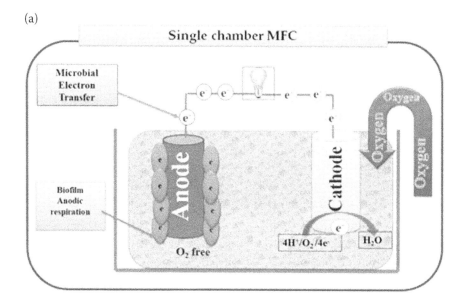

(b)

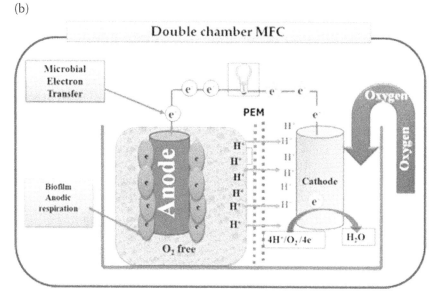

FIGURE 3.5   (a) Design of single-chamber MFC; (b) design of double chamber MFC separated by a PEM membrane. PEM stands for proton exchange membrane.

transfer the electrons, but it relies mainly on the presence of electro-active organisms such as *Shewanella* (Jain et al., 2012; Nourbakhsh et al., 2017), *Rhodoferax* (Liu et al., 2007) and *Geobacteraceae* (Holmes et al., 2004). These organisms transfer electrons directly to the anode via molecular nanowires and electrochemically active redox enzymes in their outer membrane.

Wastewater treatments coupled with clean energy production are the most important practical applications of the MFCs (Chandrasekhar et al., 2020a; Chandrasekhar et al., 2020b; Enamala et al., 2020; Logan, 2009). The power generated by MFCs depends on various factors such as the electron transfer rate from the bacteria to the anode, the diffusion of the substrate into the biofilm, the ohmic resistance of the electrolyte, and the electrochemical kinetics. The maximum power generated by the MFC also depends on the total internal resistance of the system (Cao et al., 2019; Dessie et al., 2020). The reviews of the basic MFCs technology, challenges, and applications can be found in several articles (Li et al., 2014; Wei et al., 2011).

3.6.2 ELECTROCHEMICAL BIOSENSORS

The use of biofilm in electrochemical biosensors was exploited for various detection approaches. The electrochemical biosensor systems are depending on the conductivity of biofilm and their electron transfer ability and a redox mediator could be employed. Biofilm provides direct electron transfer, called electroactive biofilm, where any redox mediators that are needed are preferred in sensing applications (Prévoteau & Rabaey, 2017). Electrochemical biosensors based on biofilms are generally used in the enzymatic detection approach exploiting catalytic reactions towards substrates. The detection is based on the use of a microorganism as a catalyst that contains different enzymes, able to (co)metabolize a broad range of substrates via multiple reaction pathways. However, the biosensor based on catalytic biofilms has less selectivity and is not designed to detect specific substrates but is attractive for monitoring a range of substrates. Thus, the proteins present in biofilm could make chemical reactions and produce an electrochemical signal. The successive enzymatic reaction in the cell improves the electron transfer and the electrochemical response is higher compared to the single enzymatic biosensor. In addition, the biosensors provide high stability and low cost compared to enzymatic biosensors, which need various steps of purification and suffer from low stability in complex media. The electron transfer could be obtained by redox mediator or directly by electroactive biofilm when direct electron transfer could be obtained.

Biosensors based on biofilm could be used in various sensing applications. For example, biosensors based on microbial biofilm were used for the detection of biodegradable organic compounds and monitor wastewater. They have been also reported to efficiently monitor various compounds such as carbohydrates, organic or amino acids, alcohols and phenols, hydrocarbons, peptides, vitamins, antibiotics, and organic or inorganic N-, S-, or P- compounds (Riedel et al., 1989; Su et al., 2011).

Microbial biosensors have also been developed for the online monitoring of the dynamic changes and cellular responses to toxic heavy metal ions. For example, the *Rhizobium-MAP7* showed high efficiency for the removal of ions of heavy metals by acquiring a strong resistivity to such toxic metals ions.

The Voltammetric currents which resulted from electrochemically active biofilm formed at the surface of the working electrodes demonstrated the differences between two microorganisms (*Rhizobium-MAP7R* and the *Rhodotorula-ALT72*) and their capacity of bioremediations. Thus, the removal of heavy metals Cr(VI) and Cd(II) could be determined (Alfadaly et al., 2021). Using the CV, the anti-biofilm agents (compounds that are used for prevention or inhibition of biofilm formation) were explored for electrochemical biosensors. Formation of *S. aureus* biofilms on electrode surfaces was constructed and the electrochemical signal of biofilm was measured before and after treatment with antibiotics or anti-biofilms such as fluoroaryl-2,2′-Bichalcophene. The electrochemical signal was highly sensitive to detect the changes in the electron transfer rates which represents the viability of microbial cells inside the biofilm matrix (Elmogy et al., 2020).

Electrochemical biosensors based on biofilm could be monitored by various bioelectrochemical systems for measuring specific compounds and following the reaction. The readout could be obtained by a microbial biofuel cell or microbial three-electrode cells.

Microbial biofuel cells have the advantage compared to three-electrode cells to deliver small electrical power that could in some cases provide for their own energy needs (autonomous biosensors). Three-electrode cells have the advantage of using a reference electrode and external electronic control for applying appropriate potential, which leads to a stable baseline current and the electrical signal completely dependent on the microbial process. The application of biofilm in biosensor devices is now well documented in literature data and various studies have reported using electroactive biofilm in sensing applications are described. The advantages are the ability for monitoring for a long period of time compared to others sensors. However, the various challenge is remaining for the application in real conditions where improvement of the nature of the electrode surface and topography where the biofilm growth is of great importance. Also, current trends in biofilm bioelectrochemistry are moving to genetic manipulation to enhance the properties of electroactive biofilms and increase the current density of microbial electrodes.

3.7 CONCLUSION

In the last 10 years, the study of biofilm and processes in bioelectrochemical systems has progressed significantly. However, there is currently a scarcity of information on the description of the whole biofilm story from the formation and characterization to the description of the mechanisms of electron transfer within biofilms and their application in bioelectrochemical systems. This chapter covers all of these items from the phases of biofilm formation to bioelectrochemistry of biofilm and the electron exchanges measurement through direct and indirect methods. The application of microbial biofilm in biofuel cells and electrochemical detection are introduced.

REFERENCES

Alfadaly, R. A., Elsayed, A., Hassan, R. Y. A., Noureldeen, A., & Darwish, H., & Gebreil, A. S. (2021). Microbial sensing and removal of heavy metals: Bioelectrochemical detection and removal of Chromium(VI) and Cadmium(II). *Molecules, 26*(9), 2549.

Azeredo, J., Azevedo, N. F., Briandet, R., Cerca, N., Coenye, T., Costa, A. R., Desvaux, M., Di Bonaventura, G., Hébraud, M., & Jaglic, Z. (2017a). Critical review on biofilm methods. *Critical Reviews in Microbiology, 43*(3), 313–351.

Azeredo, J., Azevedo, N. F., Briandet, R., Cerca, N., Coenye, T., Costa, A. R., Desvaux, M., Di Bonaventura, G., Hébraud, M., Jaglic, Z., Kačániová, M., Knøchel, S., Lourenço, A., Mergulhão, F., Meyer, R.L., Nychas, G., Simões, M., Tresse, O., & Sternberg, C. (2017b). Critical review on biofilm methods. *Critical Reviews in Microbiology, 43*(3), 313–351.

Babauta, J., Renslow, R., Lewandowski, Z., & Beyenal, H. (2012). Electrochemically active biofilms: Facts and fiction. A review. *Biofouling, 28*(8), 789–812.

Barrière, F., & Downard, A. J. (2008). Covalent modification of graphitic carbon substrates by non-electrochemical methods. *Journal of Solid State Electrochemistry, 12*(10), 1231–1244.

Barsoumian, A. E., Mende, K., Sanchez, C. J., Beckius, M. L., Wenke, J. C., Murray, C. K., & Akers, K. S. (2015a). Clinical infectious outcomes associated with biofilm-related bacterial infections: A retrospective chart review. *BMC infectious diseases, 15*(1), 1–7.

Barsoumian, A. E., Mende, K., Sanchez, C. J., Beckius, M. L., Wenke, J. C., Murray, C. K., & Akers, K. S. (2015b). Clinical infectious outcomes associated with biofilm-related bacterial infections: A retrospective chart review. *BMC Infectious Diseases, 15*(1), 223.

Baselga, R., Albizu, I., & Amorena, B. (1994). *Staphylococcus aureus* capsule and slime as virulence factors in ruminant mastitis. A review. *Veterinary Microbiology, 39*(3–4), 195–204.

Berne, C., Ducret, A., Hardy, G. G., & Brun, Y. V. (2015). Adhesins involved in attachment to abiotic surfaces by Gram-negative bacteria. *Microbial Biofilms, 3*, 163–199.

Bjarnsholt, T. (2013). The role of bacterial biofilms in chronic infections. *Apmis, 121*, 1–58.

Butler, C., & Boltz, J. (2014). Biofilm processes and control in water and wastewater treatment. In *Remediation of Polluted Waters* (Vol. 3, pp. 90–107). Elsevier Inc. https://doi.org/10.1016/B978-0-12-382182-9.00083-9

Büttner, H., Mack, D., & Rohde, H. (2015). Structural basis of *Staphylococcus epidermidis* biofilm formation: Mechanisms and molecular interactions. *Frontiers in Cellular and Infection Microbiology, 5*, 14.

Cao, Y., Mu, H., Liu, W., Zhang, R., Guo, J., Xian, M., & Liu, H. (2019). Electricigens in the anode of microbial fuel cells: Pure cultures versus mixed communities. *Microbial Cell Factories, 18*(1), 39.

Chandrasekhar, K. (2019). Effective and nonprecious cathode catalysts for oxygen reduction reaction in microbial fuel cells. In: S. V. Mohan, S. Varjani, & A. Pandey (Eds.), *Microbial electrochemical technology* (pp. 485–501). Elsevier.

Chandrasekhar, K., & Ahn, Y. H. (2017). Effectiveness of piggery waste treatment using microbial fuel cells coupled with elutriated-phased acid fermentation. *Bioresource Technology, 244*(Pt 1), 650–657.

Chandrasekhar, K., Amulya, K., & Mohan, S. V. (2015a). Solid phase bio-electrofermentation of food waste to harvest value-added products associated with waste remediation. *Waste Management, 45*, 57–65.

Chandrasekhar, K., Kadier, A., Kumar, G., Nastro, R. A., & Jeevitha, V. (2018). Challenges in microbial fuel cell and future scope. In: D. Das (Ed.), *Microbial fuel cell* (pp. 483–499). Springer International Publishing, Cham.

Chandrasekhar, K., Kumar, A. N., Raj, T., Kumar, G., & Kim, S.-H. (2021a). Bioelectrochemical system-mediated waste valorization. *Systems Microbiology and Biomanufacturing*, *1*, 432–443.

Chandrasekhar, K., Kumar, G., Venkata Mohan, S., Pandey, A., Jeon, B.-H., Jang, M., & Kim, S.H. (2020a). Microbial electro-remediation (MER) of hazardous waste in aid of sustainable energy generation and resource recovery. *Environmental Technology & Innovation*, *19*, 100997.

Chandrasekhar, K., Kumar, S., Lee, B. D., & Kim, S. H. (2020b). Waste based hydrogen production for circular bioeconomy: Current status and future directions. *Bioresource Technology*, *302*, 122920.

Chandrasekhar, K., Lee, Y. J., & Lee, D. W. (2015b). Biohydrogen production: Strategies to improve process efficiency through microbial routes. *International Journal of Molecular Sciences*, *16*(4), 8266–8293.

Chandrasekhar, K., Naresh Kumar, A., Kumar, G., Kim, D.-H., Song, Y.-C., & Kim, S.-H. (2021b). Electro-fermentation for biofuels and biochemicals production: Current status and future directions. *Bioresource Technology*, *323*, 124598.

Chandrasekhar, K., Velvizhi, G., & Venkata Mohan, S. (2021c). Bio-electrocatalytic remediation of hydrocarbons contaminated soil with integrated natural attenuation and chemical oxidant. *Chemosphere*, *280*, 130649.

Chandrasekhar, K., & Venkata Mohan, S. (2012). Bio-electrochemical remediation of real field petroleum sludge as an electron donor with simultaneous power generation facilitates biotransformation of PAH: Effect of substrate concentration. *Bioresource Technology*, *110*, 517–525.

Chandrasekhar, K., & Venkata Mohan, S. (2014a). Bio-electrohydrolysis as a pretreatment strategy to catabolize complex food waste in closed circuitry: Function of electron flux to enhance acidogenic biohydrogen production. *International Journal of Hydrogen Energy*, *39*(22), 11411–11422.

Chandrasekhar, K., & Venkata Mohan, S. (2014b). Induced catabolic bio-electrohydrolysis of complex food waste by regulating external resistance for enhancing acidogenic bio-hydrogen production. *Bioresource Technology*, *165*, 372–382.

Chao, Y., Marks, L. R., Pettigrew, M. M., & Hakansson, A. P. (2015). Streptococcus pneumoniae biofilm formation and dispersion during colonization and disease. *Frontiers in Cellular and Infection Microbiology*, *4*, 194.

Chaturongkasumrit, Y., Takahashi, H., Keeratipibul, S., Kuda, T., & Kimura, B. (2011). The effect of polyesterurethane belt surface roughness on *Listeria monocytogenes* biofilm formation and its cleaning efficiency. *Food Control*, *22*(12), 1893–1899.

Chmielewski, R., & Frank, J. (2003). Biofilm formation and control in food processing facilities. *Comprehensive Reviews in Food Science and Food Safety*, *2*(1), 22–32.

Christensen, G. D., Simpson, W. A., Anglen, J. O., & Gainor, B. J. (2000). Methods for evaluating attached bacteria and biofilms. In: Y. H. An & R. J. Friedman, *Handbook of Bacterial Adhesion*, Springer, pp. 213–233.

Christensen, G. D., Simpson, W. A., Younger, J., Baddour, L., Barrett, F., Melton, D., & Beachey, E. (1985). Adherence of coagulase-negative staphylococci to plastic tissue culture plates: A quantitative model for the adherence of staphylococci to medical devices. *Journal of Clinical Microbiology*, *22*(6), 996–1006.

Cohen, B. (1931). The bacterial culture as an electrical half-cell. *Journal of Bacteriology*, *21*(1), 18–19.

Cornejo, J. A., Lopez, C., Babanova, S., Santoro, C., Artyushkova, K., Ista, L., Schuler, A. J., & Atanassov, P. (2015). Surface modification for enhanced biofilm formation and electron transport in Shewanella anodes. *Journal of the Electrochemical Society*, *162*(9), H597–H603.

Costerton, J., Ingram, J., & Cheng, K. (1974). Structure and function of the cell envelope of gram-negative bacteria. *Bacteriological Reviews*, *38*(1), 87.

Cui, S., Li, M., Hassan, R. Y. A., Heintz-Buschart, A., Wang, J., & Bilitewski, U. (2020). Inhibition of respiration of *Candida albicans* by small molecules increases phagocytosis efficacy by macrophages. *mSphere*, *5*(2), e00016–e00020.

Davies, D. G., & Marques, C. N. (2009). A fatty acid messenger is responsible for inducing dispersion in microbial biofilms. *Journal of Bacteriology*, *191*(5), 1393–1403.

Dessie, Y., Tadesse, S., & Eswaramoorthy, R. (2020). Review on manganese oxide based biocatalyst in microbial fuel cell: Nanocomposite approach. *Materials Science for Energy Technologies*, *3*, 136–149.

Deval, A. S., Parikh, H. A., Kadier, A., Chandrasekhar, K., Bhagwat, A. M., & Dikshit, A. K. (2017). Sequential microbial activities mediated bioelectricity production from distillery wastewater using bio-electrochemical system with simultaneous waste remediation. *International Journal of Hydrogen Energy*, *42*(2), 1130–1141.

Dolch, K., Danzer, J., Kabbeck, T., Bierer, B., Erben, J., Förster, A. H., Maisch, J., Nick, P., Kerzenmacher, S., & Gescher, J. (2014). Characterization of microbial current production as a function of microbe–electrode-interaction. *Bioresource Technology*, *157*, 284–292.

Dong, X., & Stams, A. J. (1995). Evidence for H_2 and formate formation during syntrophic butyrate and propionate degradation. *Anaerobe*, *1*(1), 35–39.

Donlan, R. M., & Costerton, J. W. (2002). Biofilms: Survival mechanisms of clinically relevant microorganisms. *Clinical Microbiology Reviews*, *15*(2), 167–193.

Dufour, D., Leung, V., & Lévesque, C. M. (2010). Bacterial biofilm: Structure, function, and antimicrobial resistance. *Endodontic Topics*, *22*(1), 2–16.

Egerton, R. F. (2005). *Physical principles of electron microscopy*. Springer.

Elmogy, S., Ismail, M. A., Hassan, R. Y. A., Noureldeen, A., Darwish, H., Fayad, E., Elsaid, F., & Elsayed, A. (2020). Biological Insights of fluoroaryl-2,2'-bichalcophene compounds on multi-drug resistant *Staphylococcus aureus*. *Molecules (Basel, Switzerland)*, *26*(1), 139.

Enamala, M. K., Dixit, R., Tangellapally, A., Singh, M., Dinakarrao, S. M. P., Chavali, M., Pamanji, S. R., Ashokkumar, V., Kadier, A., & Chandrasekhar, K. (2020). Photosynthetic microorganisms (algae) mediated bioelectricity generation in microbial fuel cell: Concise review. *Environmental Technology & Innovation*, *19*, 100959.

Erable, B., Duţeanu, N. M., Ghangrekar, M. M., Dumas, C., & Scott, K. (2010). Application of electro-active biofilms. *Biofouling*, *26*(1), 57–71.

Fassel, T. A., & Edmiston, C. E. (2000). Evaluating adherent bacteria and biofilm using electron microscopy. In: Y. H. An & Richard J. Friedman, *Handbook of bacterial adhesion* (pp. 235–248). Springer.

Federle, M. J., Bassler, B. L. (2003). Interspecies communication in bacteria. *The Journal of clinical investigation*, *112*(9), 1291–1299.

Flemming, H.-C., & Wingender, J. (2010a). The biofilm matrix. *Nature Reviews Microbiology*, *8*(9), 623–633.

Flemming, H. C., & Wingender, J. (2010b). The biofilm matrix. *Nature Reviews Microbiology*, *8*(9), 623–633.

Floyd, K., Eberly, A., & Hadjifrangiskou, M. (2017). Adhesion of bacteria to surfaces and biofilm formation on medical devices. In: *Biofilms and implantable medical devices* (pp. 47–95). Elsevier.

Franklin, M. J., Chang, C., Akiyama, T., & Bothner, B. (2015). New technologies for studying biofilms. *Microbial Biofilms*, *3*, 1–32.

Freeman, D., Falkiner, F., & Keane, C. (1989). New method for detecting slime production by coagulase negative staphylococci. *Journal of Clinical Pathology*, *42*(8), 872–874.

Gambino, E., Chandrasekhar, K., & Nastro, R. A. (2021). SMFC as a tool for the removal of hydrocarbons and metals in the marine environment: A concise research update. *Environmental Science and Pollution Research*, *28*(24), 30436–30451.

Garrett, T. R., Bhakoo, M., & Zhang, Z. (2008). Bacterial adhesion and biofilms on surfaces. *Progress in Natural Science*, *18*(9), 1049–1056.

Gu, H., Hou, S., Yongyat, C., De Tore, S., & Ren, D. (2013). Patterned biofilm formation reveals a mechanism for structural heterogeneity in bacterial biofilms. *Langmuir*, *29*(35), 11145–11153.

Guo, K., Prévoteau, A., Patil, S. A., & Rabaey, K. (2015). Engineering electrodes for microbial electrocatalysis. *Current Opinion in Biotechnology*, *33*, 149–156.

Guo, Y., Wang, G., Zhang, H., Wen, H., & Li, W. (2020). Effects of biofilm transfer and electron mediators transfer on *Klebsiella quasipneumoniae* sp. 203 electricity generation performance in MFCs. *Biotechnology for Biofuels*, *13*(1), 162.

Gupta, P., Sarkar, S., Das, B., Bhattacharjee, S., & Tribedi, P. (2016). Biofilm, pathogenesis and prevention—A journey to break the wall: A review. *Archives of microbiology*, *198*(1), 1–15.

Haguenau, F., Hawkes, P., Hutchison, J., Satiat-Jeunemaître, B., Simon, G., & Williams, D. (2003). Key events in the history of electron microscopy. *Microscopy and Microanalysis*, *9*(2), 96.

Halan, B., Buehler, K., & Schmid, A. (2012). Biofilms as living catalysts in continuous chemical syntheses. *Trends in Biotechnology*, *30*(9), 453–465.

Hall-Stoodley, L., Costerton, J. W., & Stoodley, P. (2004). Bacterial biofilms: From the natural environment to infectious diseases. *Nature Reviews Microbiology*, *2*(2), 95–108.

Hall-Stoodley, L., & Stoodley, P. (2009). Evolving concepts in biofilm infections. *Cellular Microbiology*, *11*(7), 1034–1043.

Hassan, R. Y., & Bilitewski, U. (2013). Direct electrochemical determination of Candida albicans activity. *Biosensors and Bioelectronics*, *49*, 192–198.

Hassan, R. Y. A., Febbraio, F., & Andreescu, S. (2021). Microbial electrochemical systems: Principles, construction and biosensing applications. *Sensors (Basel, Switzerland)*, *21*(4), 1279.

Hassan, R. Y. A., Mekawy, M. M., Ramnani, P., & Mulchandani, A. (2017). Monitoring of microbial cell viability using nanostructured electrodes modified with graphene/alumina nanocomposite. *Biosensors and Bioelectronics*, *91*, 857–862.

Hassan, R. Y. A., & Wollenberger, U. (2019). Direct determination of bacterial cell viability using carbon nanotubes modified screen-printed electrodes. *Electroanalysis*, *31*(6), 1112–1117.

Henrici, A. T. (1933). Studies of freshwater bacteria: I. A direct microscopic technique. *Journal of Bacteriology*, *25*(3), 277.

Holmes, D. E., Nicoll, J. S., Bond, D. R., & Lovley, D. R. (2004). Potential role of a novel psychrotolerant member of the family Geobacteraceae, *Geopsychrobacter electrodiphilus* gen. nov., sp. nov., in electricity production by a marine sediment fuel cell. *Applied and Environmental Microbiology*, *70*(10), 6023–6030.

Jain, A., Zhang, X., Pastorella, G., Connolly, J. O., Barry, N., Woolley, R., Krishnamurthy, S., & Marsili, E. (2012). Electron transfer mechanism in *Shewanella loihica* PV-4 biofilms formed at graphite electrode. *Bioelectrochemistry*, *87*, 28–32.

Jendresen, M. D., & Glantz, P.-O. (1981). Clinical adhesiveness of selected dental materials: an in-vivo study. *Acta Odontologica Scandinavica*, *39*(1), 39–45.

Jendresen, M. D., Glantz, P.-O., Baier, R. E., & Eick, J. D. (1981). Microtopography and clinical adhesiveness of an acid etched tooth surface: An in-vivo study. *Acta Odontologica Scandinavica*, *39*(1), 47–53.

Joo, H.-S., & Otto, M. (2012). Molecular basis of in vivo biofilm formation by bacterial pathogens. *Chemistry & Biology*, *19*(12), 1503–1513.

Kaali, P., Strömberg, E., & Karlsson, S. (2011). *Prevention of biofilm associated infections and degradation of polymeric materials used in biomedical applications*. INTECH Open Access Publisher.

Kaplan, J. á. (2010). Biofilm dispersal: Mechanisms, clinical implications, and potential therapeutic uses. *Journal of Dental Research*, *89*(3), 205–218.

Karunakaran, C., Bhargava, K., & Benjamin, R. (2015). *Biosensors and bioelectronics*. Elsevier.

Katuri, K. P., Enright, A.-M., O'Flaherty, V., & Leech, D. (2012). Microbial analysis of anodic biofilm in a microbial fuel cell using slaughterhouse wastewater. *Bioelectrochemistry*, *87*, 164–171.

Katuri, K. P., Kamireddy, S., Kavanagh, P., Mohammad, A., Conghaile, P. Ó., Kumar, A., Saikaly, P. E., & Leech, D. (2020). Electroactive biofilms on surface functionalized anodes: the anode respiring behavior of a novel electroactive bacterium, *Desulfuromonas acetexigens*. *bioRxiv 2020.03.04.974261*. https://doi.org/10.1101/2 020.03.04.974261

Kempf, V. A., Trebesius, K., & Autenrieth, I. B. (2000). Fluorescent in situ hybridization allows rapid identification of microorganisms in blood cultures. *Journal of Clinical Microbiology*, *38*(2), 830–838.

Khater, D. Z., El-khatib, K. M., & Hassan, R. Y. A. (2018). Exploring the bioelectrochemical characteristics of activated sludge using cyclic voltammetry. *Applied Biochemistry and Biotechnology*, *184*(1), 92–101.

Khater, D. Z., El-Khatib, K. M., Hazaa, M. M., & Hassan, R. Y. (2015). Development of bioelectrochemical system for monitoring the biodegradation performance of activated sludge. *Applied Biochemistry and Biotechnology*, *175*(7), 3519–3530.

Kjeldsen, K. U., Schreiber, L., Thorup, C. A., Boesen, T., Bjerg, J. T., Yang, T., Dueholm, M. S., Larsen, S., Risgaard-Petersen, N., Nierychlo, M., Schmid, M., Bøggild, A., van de Vossenberg, J., Geelhoed, J. S., Meysman, F. J. R., Wagner, M., Nielsen, P. H., Nielsen, L. P., & Schramm, A. (2019). On the evolution and physiology of cable bacteria. *Proceedings of the National Academy of Sciences*, *116*(38), 19116.

Klausen, M., Aaes-Jørgensen, A., Molin, S., & Tolker-Nielsen, T. (2003). Involvement of bacterial migration in the development of complex multicellular structures in Pseudomonas aeruginosa biofilms. *Molecular Microbiology*, *50*(1), 61–68.

Kumar, A. K., Reddy, M. V., Chandrasekhar, K., Srikanth, S., & Mohan, S. V. (2012). Endocrine disruptive estrogens role in electron transfer: bio-electrochemical remediation with microbial mediated electrogenesis. *Bioresource Technology*, *104*, 547–556.

Kumar, C. G., & Anand, S. K. (1998). Significance of microbial biofilms in food industry: a review. *International Journal of Food Microbiology*, *42*(1–2), 9–27.

Kumar, P., Chandrasekhar, K., Kumari, A., Sathiyamoorthi, E., & Kim, B. (2018). Electro-fermentation in aid of bioenergy and biopolymers. *Energies*, *11*(2), 343.

LewisOscar, F., Vismaya, S., Arunkumar, M., Thajuddin, N., Dhanasekaran, D., & Nithya, C. (2016). Algal nanoparticles: Synthesis and biotechnological potentials. In: *Algae-organisms for imminent biotechnology*. IntechOpen.

Li, T., Yang, X.-L., Chen, Q.-L., Song, H.-L., He, Z., & Yang, Y.-L. (2020). Enhanced performance of microbial fuel cells with electron mediators from anthraquinone/polyphenol-abundant herbal plants. *ACS Sustainable Chemistry & Engineering*, *8*(30), 11263–11275.

Li, W.-W., Yu, H.-Q., & He, Z. (2014). Towards sustainable wastewater treatment by using microbial fuel cells-centered technologies. *Energy & Environmental Science*, *7*(3), 911–924.

Liang, R. M., Cao, Y. B., Zhou, Y. J., Xu, Y., Gao, P. H., Dai, B. D., Yang, F., Tang, H., & Jiang, Y. Y. (2010). Transcriptional response of *Candida albicans* biofilms following exposure to 2-amino-nonyl-6-methoxyl-tetralin muriate. *Acta Pharmacologica Sinica*, *31*(5), 616–628.

Liu, Z. D., Du, Z. W., Lian, J., Zhu, X. Y., Li, S. H., & Li, H. R. (2007). Improving energy accumulation of microbial fuel cells by metabolism regulation using *Rhodoferax ferrireducens* as biocatalyst. *Letters in Applied Microbiology*, *44*(4), 393–398.

Logan, B. E. (2009). Exoelectrogenic bacteria that power microbial fuel cells. *Nature Reviews Microbiology*, *7*(5), 375–381.

Magana, M., Sereti, C., Ioannidis, A., Mitchell, C. A., Ball, A. R., Magiorkinis, E., Chatzipanagiotou, S., Hamblin, M. R., Hadjifrangiskou, M., & Tegos, G. P. (2018). Options and limitations in clinical investigation of bacterial biofilms. *Clinical Microbiology Reviews*, *31*(3), e00084–16.

Magot, M., Ollivier, B., & Patel, B. K. (2000). Microbiology of petroleum reservoirs. *Antonie van Leeuwenhoek*, *77*(2), 103–116.

Mahmoud, R. H., Abdo, S. M., Samhan, F. A., Ibrahim, M. K., Ali, G. H., & Hassan, R. Y. (2020a). Biosensing of algal-photosynthetic productivity using nanostructured bioelectrochemical systems. *Journal of Chemical Technology & Biotechnology*, *95*(4), 1028–1037.

Mahmoud, R. H., Abdo, S. M., Samhan, F. A., Ibrahim, M. K., Ali, G. H., & Hassan, R. Y. A. (2020b). Biosensing of algal-photosynthetic productivity using nanostructured bioelectrochemical systems. *Journal of Chemical Technology & Biotechnology*, *95*(4), 1028–1037.

Mahmoud, R. H., Samhan, F. A., Ali, G. H., Ibrahim, M. K., & Hassan, R. Y. A. (2018). Assisting the biofilm formation of exoelectrogens using nanostructured microbial fuel cells. *Journal of Electroanalytical Chemistry*, *824*, 128–135.

Mahmoud, R. H., Samhan, F. A., Ibrahim, M. K., Ali, G. H., & Hassan, R. Y. A. (2021). Formation of electroactive biofilms derived by nanostructured anodes surfaces. *Bioprocess and Biosystems Engineering*, *44*(4), 759–768.

Mattila-Sandholm, T., & Wirtanen, G. (1992). Biofilm formation in the industry: a review. *Food reviews international*, *8*(4), 573–603.

McDougald, D., Rice, S. A., Barraud, N., Steinberg, P. D., & Kjelleberg, S. (2012). Should we stay or should we go: mechanisms and ecological consequences for biofilm dispersal. *Nature Reviews Microbiology*, *10*(1), 39–50.

Minaev, B. (2007). Electronic mechanisms of molecular oxygen activation. *Russian Chemical Reviews*, *76*(11), 989–1011.

Mohan, S. V., & Chandrasekhar, K. (2011a). Self-induced bio-potential and graphite electron accepting conditions enhances petroleum sludge degradation in bio-electrochemical system with simultaneous power generation. *Bioresource Technology, 102*(20), 9532–9541.

Mohan, S. V., & Chandrasekhar, K. (2011b). Solid phase microbial fuel cell (SMFC) for harnessing bioelectricity from composite food waste fermentation: influence of electrode assembly and buffering capacity. *Bioresource Technology, 102*(14), 7077–7085.

Ng, I. S., Hsueh, C.-C., & Chen, B.-Y. (2017). Electron transport phenomena of electroactive bacteria in microbial fuel cells: A review of *Proteus hauseri*. *Bioresources and Bioprocessing, 4*(1), 53.

Nickel, J., Ruseska, I., Wright, J., & Costerton, J. (1985). Tobramycin resistance of *Pseudomonas aeruginosa* cells growing as a biofilm on urinary catheter material. *Antimicrobial Agents and Chemotherapy, 27*(4), 619–624.

Nielsen, L. P., Risgaard-Petersen, N., Fossing, H., Christensen, P. B., & Sayama, M. (2010). Electric currents couple spatially separated biogeochemical processes in marine sediment. *Nature, 463*(7284), 1071–1074.

Nourbakhsh, F., Mohsennia, M., & Pazouki, M. (2017). Nickel oxide/carbon nanotube/ polyaniline nanocomposite as bifunctional anode catalyst for high-performance Shewanella-based dual-chamber microbial fuel cell. *Bioprocess and Biosystems Engineering, 40*(11), 1669–1677.

O'Toole, G., Kaplan, H. B., & Kolter, R. (2000). Biofilm formation as microbial development. *Annual Reviews in Microbiology, 54*(1), 49–79.

O'Toole, G. A., & Wong, G. C. (2016). Sensational biofilms: surface sensing in bacteria. *Current Opinion in Microbiology, 30*, 139–146.

Otto, M. (2013). Staphylococcal infections: mechanisms of biofilm maturation and detachment as critical determinants of pathogenicity. *Annual Review of Medicine, 64*, 175–188.

Paharik, A. E., & Horswill, A. R. (2016). The staphylococcal biofilm: Adhesins, regulation, and host response. *Virulence Mechanisms of Bacterial Pathogens, 4*, 529–566.

Pandit, S., Chandrasekhar, K., Kakarla, R., Kadier, A., & Jeevitha, V. (2017). Basic Principles of microbial fuel cell: Technical challenges and economic feasibility. In: V. C. Kalia, & P. Kumar, (Eds.), *Microbial applications Vol. 1* (pp. 165–188). Springer International Publishing, Cham.

Pandit, S., Sarode, S., & Chandrasekhar, K. (2018a). Fundamentals of bacterial biofilm: present state of art. In: V. C. Kalia, (Ed.), *Quorum sensing and its biotechnological applications* (pp. 43–60). Springer, Singapore.

Pandit, S., Sarode, S., Sargunaraj, F., & Chandrasekhar, K. (2018b). Bacterial-mediated biofouling: Fundamentals and control techniques. In: V. C. Kalia (Ed.), *Biotechnological applications of quorum sensing inhibitors* (pp. 263–284). Springer Singapore, Singapore.

Parsek, M. R., & Singh, P. K. (2003). Bacterial biofilms: An emerging link to disease pathogenesis. *Annual Reviews in Microbiology, 57*(1), 677–701.

Pasternak, G., Greenman, J., & Ieropoulos, I. (2018). Dynamic evolution of anodic biofilm when maturing under different external resistive loads in microbial fuel cells. Electrochemical perspective. *Journal of Power Sources, 400*, 392–401.

Pfeffer, C., Larsen, S., Song, J., Dong, M., Besenbacher, F., Meyer, R. L., Kjeldsen, K. U., Schreiber, L., Gorby, Y. A., & El-Naggar, M. Y. (2012). Filamentous bacteria transport electrons over centimetre distances. *Nature, 491*(7423), 218–221.

Potter, M. C. (1911). Electrical effects accompanying the decomposition of organic compounds. *Proceedings of the Royal Society of London. Series B, Containing Papers of a Biological Character, 84*(571), 260–276.

Prévoteau, A., & Rabaey, K. (2017). Electroactive biofilms for sensing: Reflections and perspectives. *ACS Sensors, 2*(8), 1072–1085.

Rabaey, K., Rodríguez, J., Blackall, L. L., Keller, J., Gross, P., Batstone, D., Verstraete, W., & Nealson, K. H. (2007). Microbial ecology meets electrochemistry: Electricity-driven and driving communities. *The ISME Journal, 1*(1), 9–18.

Redelman, C. V., Hawkins, M. A., Drumwright, F. R., Ransdell, B., Marrs, K., & Anderson, G. G. (2012). Inquiry-based examination of chemical disruption of bacterial biofilms. *Biochemistry and Molecular Biology Education, 40*(3), 191–197.

Reguera, G., Nevin, K. P., Nicoll, J. S., Covalla, S. F., Woodard, T. L., & Lovley, D. R. (2006). Biofilm and nanowire production leads to increased current in *Geobacter sulfurreducens* Fuel Cells. *Applied and Environmental Microbiology, 72*(11), 7345.

Riedel, K., Renneberg, R., Wollenberger, U., Kaiser, G., & Scheller, F. W. (1989). Microbial sensors: Fundamentals and application for process control. *Journal of Chemical Technology & Biotechnology, 44*(2), 85–106.

Rinaldi, A., Mecheri, B., Garavaglia, V., Licoccia, S., Di Nardo, P., & Traversa, E. (2008). Engineering materials and biology to boost performance of microbial fuel cells: A critical review. *Energy & Environmental Science, 1*(4), 417–429.

Risgaard-Petersen, N., Revil, A., Meister, P., & Nielsen, L. P. (2012). Sulfur, iron-, and calcium cycling associated with natural electric currents running through marine sediment. *Geochimica et Cosmochimica Acta, 92*, 1–13.

Ritenberg, M., Nandi, S., Kolusheva, S., Dandela, R., Meijler, M. M., & Jelinek, R. (2016). Imaging *Pseudomonas aeruginosa* biofilm extracellular polymer scaffolds with amphiphilic carbon dots. *ACS Chemical Biology, 11*(5), 1265–1270.

Roy, R., Tiwari, M., Donelli, G., & Tiwari, V. (2018). Strategies for combating bacterial biofilms: A focus on anti-biofilm agents and their mechanisms of action. *Virulence, 9*(1), 522–554.

Sabir, N., Ikram, A., Zaman, G., Satti, L., Gardezi, A., Ahmed, A., & Ahmed, P. (2017). Bacterial biofilm-based catheter-associated urinary tract infections: Causative pathogens and antibiotic resistance. *American Journal of Infection Control, 45*(10), 1101–1105.

Santoro, C., Arbizzani, C., Erable, B., & Ieropoulos, I. (2017). Microbial fuel cells: From fundamentals to applications. A review. *Journal of Power Sources, 356*, 225–244.

Saratale, G. D., Saratale, R. G., Shahid, M. K., Zhen, G., Kumar, G., Shin, H.-S., Choi, Y.-G., & Kim, S.-H. (2017a). A comprehensive overview on electro-active biofilms, role of exo-electrogens and their microbial niches in microbial fuel cells (MFCs). *Chemosphere, 178*, 534–547.

Saratale, R. G., Kuppam, C., Mudhoo, A., Saratale, G. D., Periyasamy, S., Zhen, G., Kook, L., Bakonyi, P., Nemestothy, N., & Kumar, G. (2017b). Bioelectrochemical systems using microalgae – A concise research update. *Chemosphere, 177*, 35–43.

Schink, B., & Stams, A. J. (2006). Syntrophism among prokaryotes. *The Prokaryotes, 2*, 309–335.

Sedki, M., Hassan, R. Y. A., Andreescu, S., & El-Sherbiny, I. M. (2019). Online-monitoring of biofilm formation using nanostructured electrode surfaces. *Materials Science and Engineering: C, 100*, 178–185.

Selim, H. M. M., Kamal, A. M., Ali, D. M. M., & Hassan, R. Y. A. (2017). Bioelectrochemical systems for measuring microbial cellular functions. *Electroanalysis, 29*(6), 1498–1505.

Shabani, M., Younesi, H., Pontié, M., Rahimpour, A., Rahimnejad, M., & Zinatizadeh, A. A. (2020). A critical review on recent proton exchange membranes applied in microbial fuel cells for renewable energy recovery. *Journal of Cleaner Production*, *264*, 121446.

Shunmugaperumal, T. (2010). *Biofilm eradication and prevention: a pharmaceutical approach to medical device infections*. John Wiley & Sons.

Singh, R., Paul, D., & Jain, R. K. (2006). Biofilms: Implications in bioremediation. *Trends in Microbiology*, *14*(9), 389–397.

Song, F., Koo, H., & Ren, D. (2015). Effects of material properties on bacterial adhesion and biofilm formation. *Journal of Dental Research*, *94*(8), 1027–1034.

Sousa, C., Henriques, M., & Oliveira, R. (2011). Mini-review: Antimicrobial central venous catheters – Recent advances and strategies. *Biofouling*, *27*(6), 609–620.

Srey, S., Jahid, I. K., & Ha, S.-D. (2013). Biofilm formation in food industries: A food safety concern. *Food Control*, *31*(2), 572–585.

Stams, A. J., & Plugge, C. M. (2009). Electron transfer in syntrophic communities of anaerobic bacteria and archaea. *Nature Reviews Microbiology*, *7*(8), 568–577.

Stepanović, S., Vuković, D., Hola, V., Bonaventura, G. D., Djukić, S., Ćirković, I., & Ruzicka, F. (2007). Quantification of biofilm in microtiter plates: Overview of testing conditions and practical recommendations for assessment of biofilm production by staphylococci. *Apmis*, *115*(8), 891–899.

Stoodley, P., Sauer, K., Davies, D. G., & Costerton, J. W. (2002). Biofilms as complex differentiated communities. *Annual Reviews in Microbiology*, *56*(1), 187–209.

Su, L., Jia, W., Hou, C., & Lei, Y. (2011). Review: Microbial biosensors. *Biosensors and Bioelectronics*, *26*(5), 1788–1799.

Sultana, S. T., Babauta, J. T., & Beyenal, H. (2015). Electrochemical biofilm control: A review. *Biofouling*, *31*(9–10), 745–758.

Thibane, V. S., Kock, J. L., Ells, R., van Wyk, P. W., & Pohl, C. H. (2010). Effect of marine polyunsaturated fatty acids on biofilm formation of *Candida albicans* and *Candida dubliniensis*. *Marine Drugs*, *8*(10), 2597–2604.

Veerachamy, S., Yarlagadda, T., Manivasagam, G., & Yarlagadda, P. K. (2014a). Bacterial adherence and biofilm formation on medical implants: A review. *Proceedings of the Institution of Mechanical Engineers, Part H: Journal of Engineering in Medicine*, *228*(10), 1083–1099.

Veerachamy, S., Yarlagadda, T., Manivasagam, G., & Yarlagadda, P. K. (2014b). Bacterial adherence and biofilm formation on medical implants: A review. *Proceedings of the Institution of Mechanical Engineers, Part H: Journal of Engineering in Medicine*, *228*(10), 1083–1099.

Venkata Mohan, S., Chiranjeevi, P., Chandrasekhar, K., Babu, P. S., & Sarkar, O. (2019). Acidogenic biohydrogen production from wastewater. In: A. Pandey, S. V. Mohan, J.-S. Chang, P. C. Hallenbeck, C. Larroche (Eds.), *Biohydrogen* (pp. 279–320) Elsevier.

Von Eiff, C., Jansen, B., Kohnen, W., & Becker, K. (2005). Infections associated with medical devices. *Drugs*, *65*(2), 179–214.

Wei, G., Walsh, C., Cazan, I., & Marculescu, R. (2015). Molecular tweeting: Unveiling the social network behind heterogeneous bacteria populations. Proceedings of the *6th ACM Conference on Bioinformatics, Computational Biology and Health Informatics*, pp. 366–375.

Wei, J., Liang, P., & Huang, X. (2011). Recent progress in electrodes for microbial fuel cells. *Bioresource Technology*, *102*(20), 9335–9344.

Wesolowski, J., Hassan, R. Y., Hodde, S., Bardroff, C., & Bilitewski, U. (2008). Sensing of oxygen in microtiter plates: A novel tool for screening drugs against pathogenic yeasts. *Analytical and Bioanalytical Chemistry*, *391*(5), 1731–1737.

Whitchurch, C. B., Tolker-Nielsen, T., Ragas, P. C., & Mattick, J. S. (2002). Extracellular DNA required for bacterial biofilm formation. *Science, 295*(5559), 1487.

Xu, Y., Dhaouadi, Y., Stoodley, P., & Ren, D. (2020). Sensing the unreachable: Challenges and opportunities in biofilm detection. *Current Opinion in Biotechnology, 64*, 79–84.

Yi, Y., Xie, B., Zhao, T., & Liu, H. (2018). Comparative analysis of microbial fuel cell based biosensors developed with a mixed culture and*Shewanella loihica* PV-4 and underlying biological mechanism. *Bioresource Technology, 265*, 415–421.

Zhou, G., Yuan, J., & Gao, H. (2015). Regulation of biofilm formation by BpfA, BpfD, and BpfG in *Shewanella oneidensis*. *Frontiers in Microbiology, 6*, 790.

Zobell, C. E., & Allen, E. C. (1935). The significance of marine bacteria in the fouling of submerged surfaces. *Journal of Bacteriology, 29*(3), 239.

4 Algae-Mediated Bioelectrochemical System: The Future of Algae in the Electrochemical Operations

Rehab H. Mahmoud
Water Pollution Research Department, National Research Centre (NRC), Dokki, Giza, Egypt

Hany Abd El-Raheem
Zewail City of Science and Technology, Giza, Egypt

Université Paris-Saclay, CNRS, Institut de Chimie Moléculaire et des Matériaux d'Orsay (ICMMO), Orsay, France

Rabeay Y.A. Hassan
Zewail City of Science and Technology, Giza, Egypt

Applied Organic Chemistry Department, National Research Centre (NRC), Dokki, Giza, Egypt

CONTENTS

DOI: 10.1201/9781003225430-4

4.1 INTRODUCTION

Bioelectrochemical systems have received considerable attention that use microorganisms to harvest chemical energy from organic carbon substrates to produce electrical power (Mahmoud et al., 2018; Ren 2012; Chandrasekhar et al. 2021d). For a long time, the primary emphasis has been on the exploitation of heterotrophic bacteria in these systems (Dai et al., 2021; Logan et al., 2015; Tanisho et al., 1989; Zhang et al., 2007); however, most of them are inefficient at producing electricity (Delaney et al., 1984) and they may require a high nutrition rate and an efficient electron acceptor to boost power output, but this is not cost effective (Shukla & Kumar, 2018). Thus, oxygen-producing microalgae can be used in microbial fuel cells (MFCs) instead of bacteria (Enamala et al., 2018, 2020; Saratale et al., 2017b). Lately, photoautotrophic algae have been successfully applied in several bioelectrochemical systems that operate if there isn't any additional carbon feedstock (Brayner et al., 2011; Mahmoud et al., 2020; Rosenbaum et al., 2010; Wang et al., 2014). Nevertheless, electrons are emitted in the light during photosynthesis and under the dark condition through the degradation of the carbohydrates or other carbon-based substances synthesized during photosynthesis (McCormick et al., 2015). Thus, algae-based electrochemical systems can produce energy under both the dark and light phases. Furthermore, algae's fast growth rates and capability of survival under severe environmental circumstances enhance its application in bioelectrochemistry

(Saratale et al., 2017a). One of the most common bioelectrochemical systems that have achieved great success when using algae is microbial fuel cell (MFCs) technology. Algae-based MFCs are commonly applied as bi-functional systems to restore both water and energy, thus have piqued the interest and importance of researchers (Arun et al., 2020; Yang et al., 2018). In addition, algae-based bioelectrochemical systems hold a promise in the biosensing field. Recently, microalgae is widely used in biosensor design. Over the last decade, a number of algal biosensors have been created to monitor heavy metals, herbicides, and volatile organic compounds (VOCs) (Altamirano et al., 2004). In algae-based biosensors, toxic substances in the surroundings of the cells have a significant impact on their metabolic activities, which can be converted into optical or electrical signals. Heavy metals and pesticides are the most common analytes detected by algae-based biosensors (Gosset et al., 2019; Guedri & Durrieu, 2008; Pandard et al., 1993).

4.2 PRINCIPLE OF THE ALGAE-BASED ELECTROCHEMICAL SYSTEM

4.2.1 WHERE DO ELECTRONS ORIGINATE UNDER BOTH LIGHT AND DARK CONDITIONS?

In microalgae, light energy conversion occurs through two reaction centers: photosystems 1 (PSI) and photosystems 2 (PSII), which are protein multiplexes buried in the membrane of the thylakoid (Yadavalli et al., 2020, 2021). PS1, as well as PS2, are connected and interact via a chain of electron carriers that includes enzymes and co-factors such as ferredoxin (Fd), the cytochrome (Cyt) b6f complex, plastocyanin (PC), and plastoquinone (PQ). The electron carriers are responsible for transporting electrons expelled during photosynthetic water splitting to the ultimate electron acceptor, nicotinamide adenine dinucleotide phosphate ($NADP^+$). Essentially, electrons are transferred from lower-potential electron carriers to higher-potential electron carriers. In a nutshell, chlorophyll antenna light absorption activates the PSII, which then sends its energy to P680 (the PSII primary donor). Water splitting and electron transport pathways throughout the chain (plastoquinone (PQ)/plastoquinol (PQH2) pool/cytochrome b6f (b6f)/plastocyanin (PC)/Photosystem I (PSI)/Ferredoxin (Fd)/Ferredoxin-NADP reductase (FNR)) result in NADP reduction (Figure 4.1). The Calvin cycle finally culminates in the decrease of CO_2 via sequential ATP synthesis by ATP synthase. Taking benefit from the photosynthesis process essentially occurs by using an exterior polarized electrode with high electrochemical activity for gathering a fraction of the electron flow along the photosynthetic chain (Sayegh et al., 2019). Photosystems (PSII and PSI) have been separated as photochemical converters to enhance electron transport between photosynthetic microbes and electrode surfaces in several studies (Kato et al., 2014; Sayegh et al., 2019; Tel-Vered & Willner, 2014). Nonetheless, it brings up the question of these

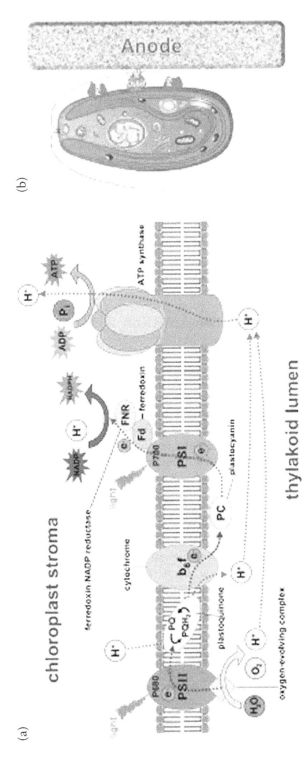

FIGURE 4.1 Schematic representation of photosynthetic electron transport. (a) Photosystem I (PSI), Photosystem II (PSII), Cytochrome b6f complex (Cyt b6f), plastocyanin (PC), Ferredoxin (Fd), and ferredoxin-NADP reductase (FNR). Cytochrome c6 (Cyt c6) transfers electrons from the Cyt b6f modified from (Simkin et al., 2019); (b) algal cell attached on the anode surface.

systems' flexibility outside of their biological environment. As a result, iso-lated thylakoid membranes or chloroplasts are also taken into account (Calkins et al., 2013; Hasan et al., 2014). However, the lack of cell proliferation in these two techniques is a significant issue that necessitates further research into entire photosynthetic organisms (Grattieri et al., 2017; Hasan et al., 2017; Sekar et al., 2014). Hence, the electron transport routes to the electrode get more complicated as the target becomes more complex. In fact, energy har-vesting directly from a photosynthetic microorganism organism is uncommon (Kim et al., 2016; Kim et al., 2018; Ryu et al., 2010). So, to boost photocurrent production, electron shuttles such as soluble mediators (quinones Fe(CN)-36), redox polymers, and nano-objects are used as supplementary agents (Sekar & Ramasamy, 2015).

In addition, during the respiration process microalgae, are able to metabolize carbohydrates and produce ATP for their internal biological processes as well as H_2O, CO_2, and electrons are regenerated (Bradley et al., 2012). The liberated electrons are migrated from the anode to the cathode through the external electrical circuit, resulting in a potential difference between electrodes (Figure 4.2) (Chandrasekhar et al., 2020b; Chandrasekhar et al., 2021a; Moriuchi et al., 2008). Eventually, protons are freed from the anodic chamber and diffuse to the cathode, where they recombine with electrons and O_2 to reform H_2O (Chandrasekhar et al., 2021c; Chandrasekhar et al., 2018; Gambino et al., 2021; Lee & Choi, 2015). In 2013, a study for isolation and characterization of electrogenic microalgae strains reported that *Desmodesmus* sp. exhibited direct electron transfer through membrane-associated proteins under dark conditions (Wu et al., 2013).

4.3 MICROALGAE-BASED MFC (MB-MFC)

In the last few decades, microbial fuel cells (MFCs), the most thoroughly char-acterized bioelectrochemical systems (BES), have made significant development. Numerous research has revealed algal-based MFCs' full potential in terms of CO_2 fixation, electricity production, and wastewater treatment versatility. In principle, there are three MB-MFC configurations (Figure 4.3): single chamber, double chamber, and algal sediment. Single-chamber MFC (Figure 4.3a) was fabricated by many researchers (Fu et al., 2009; Fu et al., 2010; Lin et al., 2013). The double-chamber MB-MFC (Figure 4.3b) is made up of cathodic and anodic chambers separated by a proton exchange membrane (PEM); in the cathodic chamber mi-croalgae is used as an oxygen supplier (Gajda et al., 2014; Venkata Mohan et al., 2014). In general, treatment of wastewater is incorporated with this configuration of MFC, in which activated sludge is incubated in the anodic chamber under anaerobic conditions (Rodrigo et al., 2009). During MFC operation, CO_2 formed by the bacterial activity of anode is transferred to the cathode chamber to be consumed by microalgae. Algal sediment MFC (Figure 4.3c) consists of an anode embedded in the sediment and a cathode compartment packed with microalgae that sit on top of the sediment. During the operation of this type of MFC for over

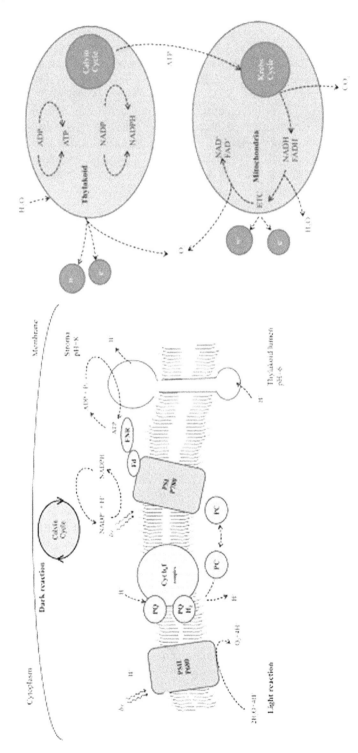

FIGURE 4.2 Illustrative diagram of electron transfer mechanism in microalgae.

Source: Adapted from Cardol et al. (2011).

(a)   (b)

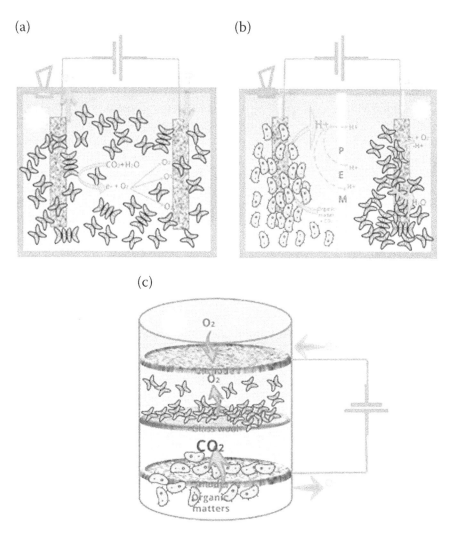

(c)

FIGURE 4.3 Schematic configurations of microalgae-based microbial fuel cell MB-MFC. (a) Single-chamber MB-MFC; (b) double-chamber MB-MFC; (c) sediment MB-MFC.

145 days, a maximum current of 0.054 ± 0.002 mA at a resistance of 1 kΩ was generated (He et al., 2009).

4.3.1 Advantages of the Algae-Based Microbial Fuel Cell (AB-MFC) Over Other Alternative Renewable Energy Sources

Through the aforementioned processes, the algae-dependent fuel cell can constantly produce electricity from light energy without adding organic substrate by improving the electrochemical potential inside the cell to split and re-create water, resulting in the production of oxygen, protons, and electrons (Rosenbaum et al., 2010). Requiring only light, CO_2, and H_2O to run algal-dependent MFC offers benefits over potentially competing for renewable energy sources, including bacterial dependent fuel cells or photovoltaic cells. Algae-based microbial fuel cells do not require an organic substrate, so there is no need for a continuous feeding system. Moreover, it is able to produce power both day and night (H. Lee & Choi, 2015), in addition to the simultaneous production of organic matter and CO_2 consumption on an electrode surface (Schenk et al., 2008) and biomass supply (Chen et al., 2012). This system represents the Earth's natural ecosystem, in which living microbes collaborate with nonliving constituents of their surroundings to provide a system that is self-sustaining and self-maintainable (H. Lee & Choi, 2015). Until now, researchers have concentrated their efforts towards demonstrating the photosynthetic electrogenic activities of different microalgae (Bradley et al., 2012; Rosenbaum et al., 2010; Strik et al., 2011).

4.3.2 FACTORS AFFECTING ALGAE-BASED MFC PERFORMANCE

Algae-based microbial fuel cell efficiency has been significantly improved by optimizing physicochemical properties, particularly in the initial stages of MFC research. The improvement in power output of MFC depends on several chemical and physical factors including temperature, light exposure, cell density and chlorophyll content, reactor design, electrode material and morphology, membrane/separator, types of substrates, pH, temperature, and the biocatalysts (Kondaveeti et al., 2017). The optimum conditions of chemical and physical parameters can improve the MFC performance by reducing internal resistance, which is the sum of overpotentials in the anode and cathode chambers, as well as the separator part, as well as by boosting Columbic efficiencies. The impact of some of these factors on algae-based MFC performance is detailed below.

4.3.2.1 The Light's Source and Intensity

The light source and intensity are crucial parameters in MA-MFCs, whether on the anodic or cathodic part because they have the potential to influence power generation by affecting algal photosynthetic activity. A previous study examined the effect of monochromatic red, and blue light of 620–750 nm as well as 450–495 nm, respectively, on the anode colonized with the green alga, *Chlamydomonas reinhardtii,* and discovered that red LED light-produced power density reached 12.95 mW m^{-2}, which was about 60% higher than the power density produced by the blue LED light at the same intensity (900 lux). This could be attributed to a faster rate of growth when *C. reinhardtii* was incubated in the red light with a wavelength ranging from 650 to 750 nm (Fu et al., 2009). Hence, it resulted in an improved photosynthesis process, electron flow rate, and power output. Furthermore, the results of the preceding study revealed that the overall chlorophyll concentration increased when light intensities increased, implying that better performance of MA-MFC can be obtained at a relatively higher light intensity. Moreover, the extreme light intensity can harm the algae-based cathodes by influencing the rate of O_2 evolution during photosynthesis. In another study, different light intensities (0, 1500, 2000, 2500, 3000, 3500 lux) were tested using cathodic *Desmodesmus*-based MFC. Light intensity was found to have a significant impact on anode and cathode resistances, and power generation. They also explored that the higher light intensity increased dissolved oxygen (DO) in the cathode biofilm, which is an indicator for increasing photosynthetic activity (Nishio et al., 2013).

4.3.2.2 Chlorophyll Content and Cell Density

Another important aspect affecting bioelectricity generation is microalgal cell density at the anode. By employing live *C. pyrenoidosa*, the highest power output was obtained at the lowest algal cell density (5.94 106 cells/mL) (Lin et al., 2013). Increased algal density has been shown to reduce terminal power output, maybe due to an increase in O_2 emissions and a mass transfer issue. Chlorophyll content of microalgae coating anode electrolyte affects power density and open-circuit voltage (OCV) (Lin et al., 2013). It was also discovered that by increasing chlorophyll content in MA-MFC up to 0.5 mg, the OCV value is increased. OCV is also related to biomass area density and is inversely proportional to light intensity. On the other side, algal concentration in the cathodic chamber of MA-MFC was found to boost power output because algae may provide more oxygen (Yifeng et al., 2011). When the biomass of green algae *Golenkinia* sp. reached a plateau during continuous operation, on the 12th day the maximum power density of 3.5 W m^{-3} was achieved and afterward, it dropped (Zhigang et al., 2019).

4.3.2.3 Dissolved Oxygen Concentration

The power output of algal MFCs can be considerably influenced by dissolved oxygen (DO) produced during photosynthesis, which lowers the Coulombic efficiency and accelerates the decomposition of aromatic compounds in the anodic chamber (Wang et al., 2013; L.-H. Yang et al., 2018). To keep the process going, the anodic chamber should be free of dissolved oxygen, or at the very least have it under control. To solve this problem, nitrogen gas or a highly concentrated salt solution has been applied (Xu et al., 2015). In this situation, microalgae should have osmotic tolerance to maintain the cell operating without compromising its physiology (Venkata Mohan et al., 2014; Khalfbadam et al., 2016).

4.3.2.4 Temperature

One of the most crucial elements impacting MFC electricity generation is temperature. MFCs can function effectively at a wide variety of temperatures. Operating temperature affects electrode potential, power density, chemical oxygen demand removal, Columbic efficiency, the internal resistance of MFCs (Oliveira et al., 2013), as well as its impact on the microbial community's composition and dispersal (González del Campo et al., 2013; Liu et al., 2011; Tang et al., 2012). Low temperatures (10°C) have a significant effect on anode performance; however, when the MFCs were run at 37°C, increasing the temperature to 43°C had a main effect on the cathode potential. The highest power density (7.89 W m^{-3}) was achieved at 37°C, while the lowest power density (2.64 W m^{-3}) was achieved at 10°C; however, increasing the temperature to 55°C indicated no consistent power generation (Li et al., 2013; Cheng et al., 2011; Michie et al., 2011; Li et al., 2013; Patil et al., 2011).

4.3.3 DIFFERENT ROLES OF MICROALGAE IN MICROBIAL FUEL CELLS

In most of the algal MFC research, live algae were never grown in an anode chamber of MFC, as photosynthesis release oxygen that is an electron acceptor and competes for electrons, leading to the reduction of electricity. For this reason, the algal role in the microbial fuel cells was limited in two functions, either at the cathode side as the oxygen provider (Juang et al., 2012; Lee et al., 2015) or at the anode side as the producer of biomass that is needed for the bacterial metabolism (Xiao & He, 2014).

4.3.3.1 Algae as Biocathode in the MFC

Because of its high oxidation potential and the fact that it produces a clean product (water) following reduction, oxygen is the most commonly employed electron acceptor in the cathode compartment (Deval et al., 2017; Flagiello et al., 2020; Nastro, 2014; Nastro et al., 2017). However, most studies demonstrate that supplying oxygen to the cathode compartment uses a lot of energy (Ucar et al., 2017). Although oxygen in the air can be directly utilized by utilizing an air cathode, the disadvantages of oxygen consumption include contact issues at the cathode-air surface and the necessity for expensive catalysts (Rismani-Yazdi et al., 2008; Ucar et al., 2017; Zhang et al., 2020). As a result, using algae as a photosynthetic biocatalyst in the cathode of an MFC is more appealing because they gather solar energy to fix CO_2 and make oxygen, which serves as the cathode's ultimate electron acceptor (Gajda et al., 2013; González del Campo et al., 2013; Powell et al., 2009). Algae in the cathodic chamber produce biomass that can be utilized as a substrate for the MFC anode, in addition to producing oxygen (Berk & Canfield, 1964; Hu et al., 2016). Green alga, *Chlorella vulgaris,* is a common cathodic alga (Song et al., 2020; Wang et al., 2010). It has already been proposed as a viable electron acceptor in the cathode of MFCs, as well as for fixing CO_2 at the same time (Powell et al., 2009). *C. vulgaris* also could be used as biocathode in the double-chambered-MFCs by cultivating *C. vulgaris* in a tubular photobioreactor as a bio-cathodic chamber. The maximum power production was more than 2.5 times higher than that of an abiotic cathode, at 21.4 mW m^2. *C. vulgaris* was also used in the cathodic chamber of MFC as a suspended or immobilized form (X. Wu et al., 2013). The immobilized form produced the most energy (2.5 W m^3), which was higher than that of the hanging form. This method also achieved an 84.8% COD reduction with zero discharge of CO_2 in a process called microbial carbon capture cell (MCC) (Zhou et al., 2012). In addition, *Chlamydomonas reinhardtii* (Nishio et al., 2013) and *Spirulina platensis* (C. Fu et al., 2010) were applied in MFCs as cathode catalysts. Table 4.1 summarizes the performance of different algae-based MFCs (Figure 4.4 and Table 4.2).

4.3.3.2   Algae as Bioanodes in the MFCs

However, research is using live algae in anode chambers as electron donors are rare as the photosynthesis produces oxygen, which serves as a terminal electron acceptor (i.e. anode competitor); there are three advantages for using live algae grown in the anode of MFC: 1) Without the addition of a substrate such as glucose, it is possible to produce electricity on a long-term basis; 2) more cleaner than other bacteria MFC than using wastewater in the anodic chamber; 3) a straightforward method for extracting energy from algae that does not require several pre-treatments of algal biomass. Researchers have attempted to solve the problem by injecting nitrogen gas or adding a highly concentrated salt solution (Xu et al., 2015; Shukla & Kumar, 2018). In 2013, *Spirulina platensis* shows the ability to directly shuttle electrons to the electrode, without the requirement of mediators. A membrane-less single-chamber PMFC reactor was operated with a photosynthetic bioanode (Chia Chi et al., 2012). In addition, the live green microalgae *Chlorella pyrenoidosa* inoculated in the anode of a microbial fuel cell (MFC) to serve as an electron donor by controlling the oxygen level without artificial mediator addition (Xu et al., 2015). In another work, when the live microalga *Oscillatoria agardhii* was introduced into the anodic chamber of the microbial fuel cell (MFC) under dark conditions, it served as an electron donor under and producing a power density of 26.8 mW m^{-2} (Mahmoud et al., 2020) (Figure 4.5).

TABLE 4.1
Performance Assessment of Algae MFCs for Electricity Production

Algae strain as biocathode	MFC type	Cathode materials	Maximum power density	Ref.
Scenedesmus acutus	Double chamber	Carbon cloth/Pt and CB	~400 mW m^{-3}	(Angioni et al., 2018)
Chlorella vulgaris	Double chamber	Carbon fiber cloth	3720 mW m^{-3}	(Zhang et al., 2019)
Spirulina sp.	Single and double chamber	Carbon cloth	(0.8–1 W m^{-2})	(Colombo et al., 2017)
Golenkinia sp.	Double chamber	Graphite plate	327.67 mW m^{-2}	(Huarachi-Olivera et al., 2018)
Mixed algal culture	Double chamber	Carbon fiber brush	50 mW m^{-2}	(Nguyen et al., 2017)
Chlorella sp.	Double chamber	Carbon fiber cloth (75 cm^2)	0.14 mW m^{-2}	(Hu et al., 2016)

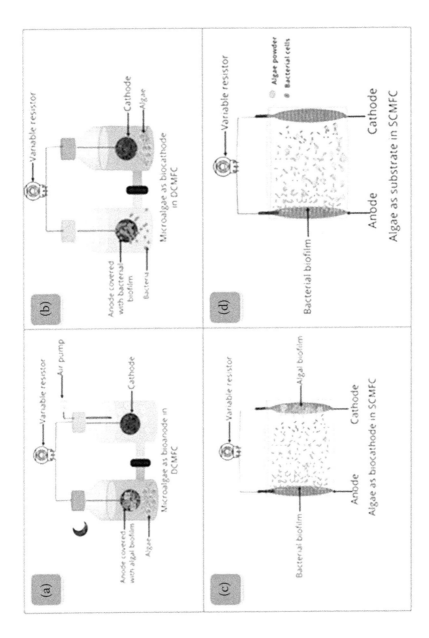

FIGURE 4.4 Schematic diagram showing the different roles of microalgae in the microbial fuel cell. (a) Microalgae as bioanode in double-chamber MFC; (b) microalgae as biocathode in double-chamber MFC; (c) microalgae as biocathode in single-chamber MFC; (d) microalgae as substrate in single-chamber MFC.

TABLE 4.2
Summary of Different Algae Used in Algal-Based Biosensors

Strain	Classification	Inorganic/organic	Detection limit	Ref
Chlorella vulgaris immobilized in BSA	Optical	Inorganic	1 ppb	(Védrine, Fabiano, et al., 2003)
Dictyosphaerium chlorelloides	Optical	Organic	0.5 µmol L^{-1}	(Altamirano et al., 2004)
Synechococcus PCC 7942 immobilized in PVA-SbQ	Optical	Inorganic/organic	0.2 and 0.06 mmol L^{-1}	(Rouillona et al., 1999)
Anabaena torulosa immobilized on an oxygen electrode	Amperometric	Inorganic		(Chay et al., 2009)
Chlorella vulgaris between two platinum electrodes	Conductometric	Inorganic	10 ppb	(Guedri & Durrieu, 2008)
Chlorella vulgaris in alginate gel	Amperometric	Organic	2–3,000 µmol dm^{-3}	(Shitanda et al., 2005b)

(a) (b)

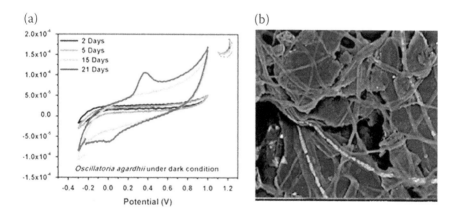

FIGURE 4.5 (a) Cyclic voltammetry of microalga *Oscillatoria agardhii* under the dark condition on the MWCNTs/MnO$_2$ modified CPE for 21 days of incubation in a closed electrochemical cell; (b) SEM images of the O agardhii biofilm formed on the anode surface.

Source: Adapted from Mahmoud et al. (2020).

4.3.3.3   Algae as Substrates in the MFCs

According to many types of research, algal biomass has an adequately high content of carbohydrates, proteins, and lipids (Liu & Cheng, 2014). The use of microalgae biomass as a substrate for MFC showed double benefits, including pollution reduction and cost-effective feedstock in MFC processing. Some microalgae have extremely high cellulose and hemicellulose content, and pre-treatment of the algal biomass is frequently required to improve process efficiency. Algal biomass either in the form of dry biomass or living cells can be used as an anodic substrate (Salar-García et al., 2016; Xiao & He, 2014). *Chlorella vulgaris* biomass has high nutritional value comprising of carbohydrates (12–55%) and proteins (42–55%) that can be mineralized by electrogenic bacteria to generate electricity (Cui et al., 2014; Safi et al., 2014). In a previous study, *Chlorella vulgaris* and *Ulva lactuca* feedstocks have tested in dry powder as a substrate for MFC operation, the power density produced was 0.98 Wm-^{2} from the *Chlorella vulgaris,* and 760 mWm^{-2} for *Ulva lactuca*–based MFC (Velasquez-Orta et al., 2009). In addition, the lipid-extracted algal (LEA) biomass was then used as an electron donor substrate in MFCs, producing a power density of 2.7 Wm^{-3} that was 145% higher than fruit waste fed MFCs (FP-MFCs) (Khandelwal et al., 2018). Furthermore, *Scenedesmus* algal biomass was tested as a nutrient source at the anode; in this work, it was observed that sonication and thermal pre-treatment of algal biomass had enhanced the microbial digestibility of the algae and also increased the overall MFC performance (Rashid et al., 2013).

4.3.4   DIFFERENT APPLICATIONS OF ALGAE-BASED MFCS

Apart from power output, algae-based MFCs have shown theirpotential in several other dimensions like wastewater treatment, CO_2 sequestration, heavy metal removal, bioremediation, biohydrogen production, biosensors, water desalination, etc. Some of these potential applications are summarized below.

4.3.4.1   Biofuel Production

The algal biomass produced in the MFCs' anode and/or cathode chambers could be used to make a variety of biofuels. Some algal species have high fatty acid content and can be used to make biodiesel, whereas carbohydrate-rich algal species can be used to make bioethanol, biogas, biohydrogen, and other biofuels through microbial conversion (Figure 4.6).

4.3.4.1.1   Biohydrogen

Hydrogen production from an algal biomass can be done in a variety of ways, including photobiological and fermentative methods (Shaishav et al., 2013). *Chlamydomonas reinhardtii* and *Scenedesmus obliquus* are the most studied microalgae to produce biohydrogen (Zhiman et al., 2011). The major enzyme that catalyzes these reactions is hydrogenase. In a photobiological reaction, ferredoxin is oxidized by the hydrogenase enzyme in the electron transfer chain,

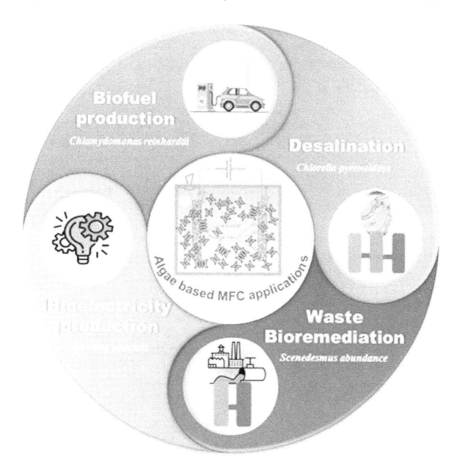

FIGURE 4.6 Different applications of algae-based microbial fuel cell.

releasing hydrogen. In dark fermentation, hydrolysis and acidogenesis are carried out by hydrogen-producing bacteria such as *Clostridium* sp., *Enterobacter* sp., *Lactobacillus* sp., *Bacillus* sp., *Klebsiella* sp., and *Citrobacter* sp. (Kawagoshi et al., 2005; Rossi et al., 2012). Genetic alterations are being used to modify hydrogen generation pathways in algae in order to boost biohydrogen yield (Batyrova & Hallenbeck, 2017; Saifuddin & Priatharsini, 2016).

4.3.4.1.2 Biogas
Specific microorganisms may break down algal organic molecules under anaerobic conditions to create methane and carbon dioxide. An algal biomass can be used to produce biogas without the need for biomass harvesting, dewatering, drying, or oil extraction. Alternatively, after extracting oil, the remaining biomass can be digested to reduce waste while still generating energy. Biomethane is widely utilized as a fuel and as a raw material for other compounds like

methanol (Chandrasekhar, Cayetano, et al., 2020a; Chandrasekhar, Mehrez, et al., 2021b). In comparison to other feedstocks, marine algae offer a greater potential for biomethanation (Chynoweth et al., 1993).

4.3.4.1.3 Bioethanol

Because the cell wall of microalgae is free of lignin and has a thin cellulose wall, it does not require extensive pre-treatment prior to fermentation (Mendez et al., 2014) carbohydrate captured in the cell. Fermenting de-oiled algal biomass is a good way to avoid the pre-treatment phase. The most common methanologen is *Saccharomyces cerevisiae;* however, it is incapable of using pentoses (Markou et al., 2013; Sulfahri et al., 2011; Kremer et al., 2015). Nitrogen fixation in the space fermentor will shorten the fermentation time (Sulfahri et al., 2011; Sulfahri Amin et al., 2016). To turn sugars into bioethanol, either fermentation or a gasification method can be used (Ullah et al., 2014). Ethanol, acetate, hydrogen, and carbon dioxide are all results of microalgal fermentation.

4.3.4.1.4 Biodiesel

Using wet algal biomass straight in a one-step process cuts down on time and solvent costs, making the method more cost effective (P. D. Patil et al., 2012). The removal of the extraction stage indirect or wet transesterification essentially means using the entire biomass as a fuel for the reaction. Surprisingly, skipping the extraction phase increased efficiency, resulting in a higher lipid output per gram of biomass. To avoid the high costs associated with dewatering and drying of algae, thermal liquefaction (Odhong et al., 2019) or hydrogenolysis (to produce hydrogen) are also promising.

4.3.4.2   Waste Bioremediation

Various studies have shown that microalgae can be used in the bioremediation process (Abo-Shady et al., 2017; Essa et al., 2018). Microalgae could be used in wastewater treatment by connecting a MFC to a photobioreactor (PBR), which exchanges the released gases. Indeed, using such toxic and polluted substrates is a huge benefit in and of itself. Several wastes and wastewaters, including sewage sludge and urine, could be used as the substrate for MFCs (Gajda et al., 2018; Ieropoulos et al., 2012), domestic wastewater (Jiang et al., 2013), food waste (Zhigang et al., 2019), brewery wastes (da Costa, 2018), pharmaceutical wastewater (Yeruva et al., 2016), and petroleum hydrocarbon (Hou et al., 2019) (Nayak & Ghosh, 2020). The treatment of kitchen wastewater was reported using Ma-MFC consisting of living *Synechococcus* sp. and *Chlorococcum* sp. as a cathode catalyst. The highest power density (41.5 mWm^{-2}) was recorded with *Synechococcus* sp. compared to *Chlorococcum* sp. (30.2 mWm^{-2}). These integrated systems combine the benefits of continuous biotreatment of various types of wastewater with the production of power and microalgal biomass while having no negative impact on the environment (Mohamed et al., 2020).

4.3.4.3 Desalination

Microalgae demonstrated significant advancements in the desalination technology in special fuel cells called microalgae assisted microbial desalination cell (MA-MDC). MA-MDCs is a new, green, and eco-friendly approach that efficiently desalinates seawater along with generating bioelectricity. In comparison to traditional desalination techniques, which require high-grade electricity and have high operating costs, this integrated approach can reduce greenhouse gas emissions as well as waste heat emissions. The idea of MA-MDC is similar to that of a typical algae-based MFC. By inserting a third chamber between the two electrodes of a standard MFC and filling it with saltwater, the MFC's cathode and anode attract the positive and negative salt ions present in the water, respectively. The semi-permeable membranes between chambers filter out the salt from the seawater, as seen in Figure 4.7 (Neethu et al., 2019; Ashwaniy & Perumalsamy, 2017; Kokabian & Gude, 2013). Microalgae performance is enhanced during natural light/dark cycles, regardless of COD concentrations, according to the findings. Based on these data, it is conceivable to conclude that algal-assisted MDC can provide a renewable, promising, and long-term wastewater desalination technique that also has the potential to generate electricity and algal biomass.

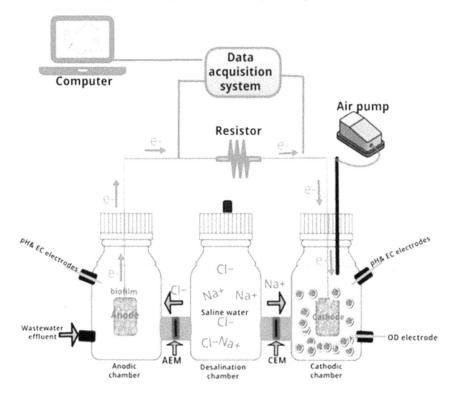

FIGURE 4.7   Schematic illustration of microalgae assisted microbial desalination cell (MA-MDC).

4.4 ALGAE-BASED BIOSENSORS

Algae have been extensively used in biosensing applications, in addition to biomass generation and bioremediation. Algae-based biosensors have shown promise in detecting analytes of agro-environmental and security concerns in a sensitive, long-term, and multiplexed manner. Over the last decade, several algal biosensors have been created to detect herbicides, volatile organic compounds (VOC), heavy metals, and even chemical warfare agents (Altamirano et al., 2004). In algae-based biosensors, the metabolic activities of the living micro-organism are evaluated. Toxic compounds in the cells' environment have a significant impact on their metabolic activity, which can be translated into electrical or optical signals. Pesticides, a broad term that encompasses herbicides, insecticides, and fungicides, are one of the main target analytes of microalgal biosensors. One of the most important advantages of algae-based biosensors is that they allow for frequent readings without the need for considerable sample preparation. Nonetheless, they usually have low selectivity in separate analyses. There is only one global signal obtained that represents a variety of toxins. However, for assessing water quality, this global response is frequently more useful than individual concentration measurements.

4.4.1 HISTORY OF ALGAE-BASED BIOSENSORS

Algae-based biosensors are a fascinating and relatively recent biotechnological issue. In the 1990s, algal biosensing systems were first used. These were distinguished by simple tools based on algae in a solution that could alter their relative cell growth and photosynthetic activity in the appearance of xenobiotics. By rough analysis, these were early attempts to use algae-based bioassays for monitoring environmental pollution (Turbak et al., 1986; Wong & Beaver, 1980). Following that, changes in the photosynthetic system followed by xenobiotic interactions were more accurately described, correlating cell death with events in electron transport (Jansen et al., 1993; Kless et al., 1994). This technique has also been applied to the development of second-generation algal biosensors based on complete algae cells immobilized on a Clark electrode for correlating oxygen evolution with hazardous chemical concentrations (Mingazzini et al., 1997; Naessens & Tran-Minh, 1998; Pandard & Rawson, 1993). Likewise, optical biosensors based on electron transfer blockage in terms of fluctuations in chlorophyll a fluorescence emission in the presence of contaminants were created (Frense et al., 1998; Kornet et al., 1992).

4.4.2 COMPOSITION AND MECHANISM OF ALGAE-BASED BIOSENSORS

A substantial body of evidence has indicated the argument and versatility of algae-based biosensor technology over the last several decades. By utilizing arrays of interchangeable receptors, these systems enable the simultaneous analysis of a wide spectrum of contaminants, such as herbicides (Belaïdi et al.,

2019), heavy metals (Campanella et al., 2001), and chemical weapons. Consequently, the double transduction framework can permit for the analysis of various matrices with varying physicochemical properties) without the need for time-consuming and potentially harmful sample pre-treatment. Even so, optical transduction has very high sensitivity and is an appropriate tool for herbicide assessment in drinking water that meets the requirements of EU directives on maximum residue levels (MRLs). Whole algal cells have the potential to provide extremely sensitive herbicide analysis (Antonacci & Scognamiglio, 2020). Other benefits of algae-based biosensors provide the ability to monitor continuously and device portability, which allows for in-field assessment. In the literature, fascinating examples of miniaturized interconnected devices developed to be flexible, adaptable, and light have been studied (Scognamiglio et al., 2012, 2013; Turemis et al., 2018). Recent advancements in microelectronics, nano-technology, microfluidics, rational design, as well as new supporting materials are paving the way for more competitive biosensors to be developed. As a result, current research is focusing on fixing the flaws of algae-based biosensors.

4.4.3 ADVANTAGES OF ALGAE-BASED BIOSENSORS

There are various algal receptors in algae-based biosensors, such as whole cells and their photosynthetic subcomponents, their ability to be combined into dual transduction miniature devices, and the ability to monitor the environment continuously. Despite challenges including poor selectivity and stability, algae-based biosensing is a viable option with several potential approaches. The entire potential of algae-based sensors will be realized through the strategic use of cutting-edge technologies such as nanotechnology, materials science, genome editing, and microfluidics. Just a global signal representing a variety of harmful chemicals is acquired. Yet, rather than measuring individual concentrations, this broad reaction is frequently more valuable for assessing water quality. Continuous monitoring and device portability are also advantages of algae-based biosensors, allowing for in-field study (Antonacci & Scognamiglio, 2020).

4.4.4 APPLICATIONS OF ALGAE-BASED BIOSENSORS

Aside from energy conversion, algae-based bioelectrochemical systems have the ability to detect pesticides, heavy metals, and other soluble analytes that alter photosynthetic processes.

4.4.4.1 Detection of Pollutant Herbicides

Owing to their sensitivity especially to herbicides, algal biosensors are valuable tools for real-time monitoring of contaminants and evaluating metabolic per-turbation reactions. Several researchers have already discussed the progress of algal biosensors for herbicide detection (Campanella et al., 2001; Chouteau et al., 2005). Pesticide analysis in contaminated liquid media with low detection limits has been widely studied using algal optical biosensors, depending on photosynthesis disturbance (designed to target chlorophyll fluorescence disruption) (Ferro et al., 2012; Naessens et al., 2000). The herbicide chemical family, primarily phenylureas, triazines, triazinones, and others were investigated for their influence on photosystem II activity in primary producers (by studying chlorophyll fluorescence) (Rioboo et al., 2002). The green alga *Chlorella vulgaris* is often employed to develop biosensors due to their better stability in emitting biological signals. Pesticides can be detected using the chlorophyll fluorescence emitted by its photosynthetic activity (Védrine et al., 2003).

4.4.4.2 Detection of Heavy Metals

The green alga *Dictyosphaerium chlorelloides* has been applied with an optic fiber linked to a microwell-plate reader or flow cell to monitor Cu^{2+} in reservoirs and water supplies (Peña-Vázquez et al., 2010). Similarly, electrochemical biosensing technologies for screening toxic metals (Ni^{2+}, toluene, or Cu^2) have been improved using the flagellate microalga *Chlamydomonas reinhardtii*'s motility (Shitanda et al., 2005a). For real-time detection and online monitoring, whole-cell biosensors based on enzymes (phosphatase and esterase) or chlorophyll fluorescence inhibition have been developed (Durrieu et al., 2006).

4.5 CONCLUSIONS

Algae-mediated bioelectrochemical systems are promising devices that can be used to generate a number of different bioenergy sources, including biofuels, biohydrogen, biodiesel, biomethane, and biomass, as well as for the biosensing field. The incorporation of photosynthetic systems with bioelectric processes has opened up a wide range of possibilities, including electricity production, wastewater treatment, and the generation of valuable materials (methanol and formic acid). Multiple trials have been performed in this area to investigate an extensive selection of electrode materials and substrates. Thorough testing on the final product and protein complexes is required in order to make the process financially viable. The marketing of biophotovoltaic devices/solar cells is accelerating due to their attractive features and high output. In addition, the feedstock mandated for its installation is either cyanobacteria or microalgae, both of which are plentiful to grow when food security is taken into account. The greatest feature is that these systems are self-sustaining and regenerative. Some isolates, in addition to being adapted to grow in seawater and freshwater, can also grow in wastewater. As a result, they can also be used for wastewater treatment.

REFERENCES

Abo-Shady, A., Khairy, H. M., Abomohra, A., Elshobary, M. E., & Essa, D. (2017). Influence of algal bio-treated industrial wastewater of two companies at Kafr El-Zayat city on some growth parameters of *Vicia faba*. *Egyptian Journal of Experimental Biology (Botany)*, *13*(2), 209–217.

Altamirano, M., Garcıa-Villada, L., Agrelo, M., Sánchez-Martın, L., Martın-Otero, L., Flores-Moya, A., Rico, M., López-Rodas, V., & Costas, E. (2004). A novel approach to improve specificity of algal biosensors using wild-type and resistant mutants: An application to detect TNT. *Biosensors and Bioelectronics*, *19*(10), 1319–1323.

Angioni, S., Millia, L., Mustarelli, P., Doria, E., Temporiti, M. E., Mannucci, B., Corana, F., & Quartarone, E. (2018). Photosynthetic microbial fuel cell with polybenzimidazole membrane: Synergy between bacteria and algae for wastewater removal and biorefinery. *Heliyon*, *4*(3), e00560.

Antonacci, A., & Scognamiglio, V. (2020). Biotechnological advances in the design of algae-based biosensors. *Trends in Biotechnology*, *38*(3), 334–347.

Arun, S., Sinharoy, A., Pakshirajan, K., & Lens, P. N. L. (2020). Algae based microbial fuel cells for wastewater treatment and recovery of value-added products. *Renewable and Sustainable Energy Reviews*, *132*, 110041. https://doi.org/10.1016/j.rser.2020.110041

Ashwaniy, V. R. V., & Perumalsamy, M. (2017). Reduction of organic compounds in petro-chemical industry effluent and desalination using *Scenedesmus abundans* algal microbial desalination cell. *Journal of Environmental Chemical Engineering*, *5*(6), 5961–5967.

Batyrova, K., & Hallenbeck, P. C. (2017). Hydrogen production by a *Chlamydomonas reinhardtii* strain with inducible expression of photosystem II. *International Journal of Molecular Sciences*, *18*(3), 647.

Belaïdi, F. S., Farouil, L., Salvagnac, L., Temple-Boyer, P., Séguy, I., Heully, J.-L., Alary, F., Bedel-Pereira, E., & Launay, J. (2019). Towards integrated multi-sensor platform using dual electrochemical and optical detection for on-site pollutant detection in water. *Biosensors and Bioelectronics*, *132*, 90–96.

Berk, R. S., & Canfield, J. H. (1964). Bioelectrochemical energy conversion. *Applied Microbiology*, *12*(1), 10–12.

Bradley, R. W., Bombelli, P., Rowden, S. J. L., & Howe, C. J. (2012). Biological photovoltaics: Intra-and extra-cellular electron transport by cyanobacteria. *Biochemical Society Transactions*, *40*(6), 1302–1307.

Brayner, R., Couté, A., Livage, J., Perrette, C., & Sicard, C. (2011). Micro-algal biosensors. *Analytical and Bioanalytical Chemistry*, *401*, 581–597. https://doi.org/10.1007/s00216-011-5107-z

Calkins, J. O., Umasankar, Y., O'Neill, H., & Ramasamy, R. P. (2013). High photoelectrochemical activity of thylakoid–carbon nanotube composites for photosynthetic energy conversion. *Energy & Environmental Science*, *6*(6), 1891–1900.

Campanella, L., Cubadda, F., Sammartino, M. P., & Saoncella, A. (2001). An algal biosensor for the monitoring of water toxicity in estuarine environments. *Water Research*, *35*(1), 69–76.

Cardol, P., Forti, G., & Finazzi, G. (2011). Regulation of electron transport in microalgae. *Biochimica et Biophysica Acta (BBA)-Bioenergetics*, *1807*(8), 912–918.

Chandrasekhar, K., Cayetano, R. D. A., Mehrez, I., Kumar, G., & Kim, S.-H. (2020a). Evaluation of the biochemical methane potential of different sorts of Algerian date biomass. *Environmental Technology & Innovation*, *20*, 101180. https://doi.org/10.1016/j.eti.2020.101180

Chandrasekhar, K., Kumar, G., Venkata Mohan, S., Pandey, A., Jeon, B. H., Jang, M., & Kim, S. H. (2020b). Microbial electro-remediation (MER) of hazardous waste in aid of sustainable energy generation and resource recovery. In: *Environmental Technology and Innovation* (Vol. 19, p. 100997). https://doi.org/10.1016/j.eti.2020.100997

Chandrasekhar, K., Kumar, A. N., Raj, T., Kumar, G., & Kim, S.-H. (2021a). Bioelectrochemical system-mediated waste valorization. *Systems Microbiology and Biomanufacturing 2021*(1), 1–12. https://doi.org/10.1007/S43393-021-00039-7

Chandrasekhar, K., Mehrez, I., Kumar, G., & Kim, S.-H. (2021b). Relative evaluation of acid, alkali, and hydrothermal pre-treatment influence on biochemical methane potential of date biomass. *Journal of Environmental Chemical Engineering, 9*(5), 106031. https://doi.org/10.1016/j.jece.2021.106031

Chandrasekhar, K., Naresh Kumar, A., Kumar, G., Kim, D. H., Song, Y. C., & Kim, S. H. (2021c). Electro-fermentation for biofuels and biochemicals production: Current status and future directions. *Bioresource Technology, 323* (December 2020), 124598. https://doi.org/10.1016/j.biortech.2020.124598

Chandrasekhar, K., Velvizhi, G., & Venkata Mohan, S. (2021d). Bio-electrocatalytic remediation of hydrocarbons contaminated soil with integrated natural attenuation and chemical oxidant. *Chemosphere, 280,* 130649. https://doi.org/10.1016/j.chemosphere.2021.130649

Chandrasekhar, K., Kadier, A., Kumar, G., Nastro, R. A., & Jeevitha, V. (2018). Challenges in microbial fuel cell and future scope. In: D. Das (Ed.), *Microbial fuel cell: A bioelectrochemical system that converts waste to Watts* (pp. 483–499). Springer International Publishing. https://doi.org/10.1007/978-3-319-66793-5_25

Chay, T., Salmijah, S., & Heng, L. (2009). The behavior of immobilized cyanobacteria *Anabaena torulosa* as an electrochemical toxicity biosensor. *Asian Journal of Biological Sciences, 2*(1), 14–20.

Chen, Z., Huang, Y., Liang, J., Zhao, F., & Zhu, Y. (2012). A novel sediment microbial fuel cell with a biocathode in the rice rhizosphere. *Bioresource Technology, 108,* 55–59.

Cheng, S., Xing, D., & Logan, B. E. (2011). Electricity generation of single-chamber microbial fuel cells at low temperatures. *Biosensors and Bioelectronics, 26*(5), 1913–1917.

Chia Chi, L., Wei, C.-H., Chen, C.-I., Shieh, C.-J., & Liu, Y.-C. (2012). Characteristics of the photosynthesis microbial fuel cell with a *Spirulina platensis* biofilm. *Bioresource Technology, 135,* 640–643. https://doi.org/10.1016/j.biortech.2012.09.138

Chouteau, C., Dzyadevych, S., Durrieu, C., & Chovelon, J.-M. (2005). A bi-enzymatic whole cell conductometric biosensor for heavy metal ions and pesticides detection in water samples. *Biosensors and Bioelectronics, 21*(2), 273–281.

Chynoweth, D. P., Turick, C. E., Owens, J. M., Jerger, D. E., & Peck, M. W. (1993). Biochemical methane potential of biomass and waste feedstocks. *Biomass and Bioenergy, 5*(1), 95–111.

Colombo, A., Marzorati, S., Lucchini, G., Cristiani, P., Pant, D., & Schievano, A. (2017). Assisting cultivation of photosynthetic microorganisms by microbial fuel cells to enhance nutrients recovery from wastewater. *Bioresource Technology, 237,* 240–248.

Cui, Y., Rashid, N., Hu, N., Saif, M., Rehman, U., & Han, J. (2014). Electricity generation and microalgae cultivation in microbial fuel cell using microalgae-enriched anode and bio-cathode. *Energy Conversion and Management, 79,* 674–680. https://doi.org/10.1016/j.enconman.2013.12.032

da Costa, C. (2018). Bioelectricity production from microalgae-microbial fuel cell technology (MMFC). *MATEC Web of Conferences, 156,* 1017.

Dai, H. N., Duong Nguyen, T.-A., My LE, L.-P., Tran, M. Van, Lan, T.-H., & Wang, C.-T. (2021). Power generation of *Shewanella oneidensis* MR-1 microbial fuel cells in bamboo fermentation effluent. *International Journal of Hydrogen Energy*, *46*(31), 16612–16621. https://doi.org/10.1016/j.ijhydene.2020.09.264

Delaney, G. M., Bennetto, H. P., Mason, J. R., Roller, S. D., Stirling, J. L., & Thurston, C. F. (1984). Performance of fuel cells containing selected microorganism-mediatorsubstrate combinations. *Journal of Chemical Technology and Biotechnology*, *34*, 13–27.

Deval, A. S., Parikh, H. A., Kadier, A., Chandrasekhar, K., Bhagwat, A. M., & Dikshit, A. K. (2017). Sequential microbial activities mediated bioelectricity production from distillery wastewater using bio-electrochemical system with simultaneous waste remediation. *International Journal of Hydrogen Energy*, *42*(2), 1130–1141. https://doi.org/10.1016/j.ijhydene.2016.11.114

Durrieu, C., Tran-Minh, C., Chovelon, J.-M., Barthet, L., Chouteau, C., & Védrine, C. (2006). Algal biosensors for aquatic ecosystems monitoring. *The European Physical Journal Applied Physics*, *36*(2), 205–209.

Enamala, M. K., Dixit, R., Tangellapally, A., Singh, M., Dinakarrao, S. M. P., Chavali, M., Pamanji, S. R., Ashokkumar, V., Kadier, A., & Chandrasekhar, K. (2020). Photosynthetic microorganisms (algae) mediated bioelectricity generation in microbial fuel cell: Concise review. *Environmental Technology & Innovation*, *19*, 100959. https://doi.org/10.1016/j.eti.2020.100959

Enamala, M. K., Enamala, S., Chavali, M., Donepudi, J., Yadavalli, R., Kolapalli, B., Aradhyula, T. V., Velpuri, J., & Kuppam, C. (2018). Production of biofuels from microalgae – A review on cultivation, harvesting, lipid extraction, and numerous applications of microalgae. *Renewable and Sustainable Energy Reviews*, *94*(May), 49–68. https://doi.org/10.1016/j.rser.2018.05.012

Essa, D., Abo-Shady, A., Khairy, H., Abomohra, A. E.-F., & Elshobary, M. (2018). Potential cultivation of halophilic oleaginous microalgae on industrial wastewater. *Egyptian Journal of Botany*, *58*(2), 205–216.

Ferro, Y., Perullini, M., Jobbagy, M., Bilmes, S. A., & Durrieu, C. (2012). Development of a biosensor for environmental monitoring based on microalgae immobilized in silica hydrogels. *Sensors*, *12*(12), 16879–16891. https://doi.org/10.3390/s121216879

Flagiello, F., Gambino, E., Nastro, R. A., & Kuppam, C. (2020). Harvesting energy using compost as a source of carbon and electrogenic bacteria. In: P. Kumar & C. Kuppam, *Bioelectrochemical Systems* (pp. 217–234). Springer Singapore. https://doi.org/10.1007/978-981-15-6868-8_9

Frense, D., Müller, A., & Beckmann, D. (1998). Detection of environmental pollutants using optical biosensor with immobilized algae cells. *Sensors and Actuators B: Chemical*, *51*(1–3), 256–260.

Fu, C.-C., Su, C.-H., Hung, T.-C., Hsieh, C.-H., Suryani, D., & Wu, W.-T. (2009). Effects of biomass weight and light intensity on the performance of photosynthetic microbial fuel cells with *Spirulina platensis*. *Bioresource Technology*, *100*(18), 4183–4186.

Fu, C., Hung, T., Wu, W., Wen, T., & Su, C. (2010). Current and voltage responses in instant photosynthetic microbial cells with Spirulina platensis. *Biochemical Engineering Journal*, *52*(2–3), 175–180. https://doi.org/10.1016/j.bej.2010.08.004

Gajda, I., Greenman, J., Melhuish, C., & Ieropoulos, I. (2013). Photosynthetic cathodes for Microbial Fuel Cells. *International Journal of Hydrogen Energy*, *38*, 1–6. https://doi.org/10.1016/j.ijhydene.2013.02.111.

Gajda, I., Greenman, J., Santoro, C., Serov, A., Melhuish, C., Atanassov, P., & Ieropoulos, I. A. (2018). Improved power and long term performance of microbial fuel cell with Fe-NC catalyst in air-breathing cathode. *Energy*, *144*, 1073–1079.

Gajda, I., Stinchcombe, A., Greenman, J., Melhuish, C., & Ieropoulos, I. (2014). Algal 'lagoon' effect for oxygenating MFC cathodes. *International Journal of Hydrogen Energy, 39*(36), 21857–21863.

Gambino, E., Chandrasekhar, K., & Nastro, R. A. (2021). SMFC as a tool for the removal of hydrocarbons and metals in the marine environment: A concise research update. *Environmental Science and Pollution Research, 28*,1–16. https://doi.org/10.1007/s11356-021-13593-3.

González del Campo, A., Cañizares, P., Rodrigo, M. A., Fernández, F. J., & Lobato, J. (2013). Microbial fuel cell with an algae-assisted cathode: A preliminary assessment. *Journal of Power Sources, 242,* 638–645. https://doi.org/10.1016/j.jpowsour.2013.05.110

Gosset, A., Oestreicher, V., Perullini, M., Bilmes, S. A., Jobbágy, M., Dulhoste, S., Bayard, R., & Durrieu, C. (2019). Optimization of sensors based on encapsulated algae for pesticide detection in water. *Analytical Methods, 11*(48), 6193–6203.

Grattieri, M., Hasan, K., & Minteer, S. D. (2017). Bioelectrochemical systems as a multi-purpose biosensing tool: Present perspective and future outlook. *ChemElectroChem, 4*(4), 834–842.

Guedri, H., & Durrieu, C. (2008). A self-assembled monolayers based conductometric algal whole cell biosensor for water monitoring. *Microchimica Acta, 163*(3–4), 179–184.

Hasan, K., Dilgin, Y., Emek, S. C., Tavahodi, M., Åkerlund, H., Albertsson, P., & Gorton, L. (2014). Photoelectrochemical communication between thylakoid membranes and gold electrodes through different quinone derivatives. *ChemElectroChem, 1*(1), 131–139.

Hasan, K., Grippo, V., Sperling, E., Packer, M. A., Leech, D., & Gorton, L. (2017). Evaluation of photocurrent generation from different photosynthetic organisms. *ChemElectroChem, 4*(2), 412–417.

He, Z., Kan, J., Mansfeld, F., Angenent, L. T., & Nealson, K. H. (2009). Self-sustained phototrophic microbial fuel cells based on the synergistic cooperation between photosynthetic microorganisms and heterotrophic bacteria. *Environmental Science and Technology, 43*(5), 1648–1654. https://doi.org/10.1021/es803084a

Hou, B., Lu, J., Wang, H., Li, Y., Liu, P., Liu, Y., & Chen, J. (2019). Performance of microbial fuel cells based on the operational parameters of biocathode during simultaneous Congo red decolorization and electricity generation. *Bioelectrochemistry, 128,* 291–297.

Hu, X., Zhou, J., & Liu, B. (2016). Effect of algal species and light intensity on the performance of an air-lift-type microbial carbon capture cell with an algae-assisted cathode. *RSC Advances, 6*(30), 25094–25100.

Huarachi-Olivera, R., Dueñas-Gonza, A., Yapo-Pari, U., Vega, P., Romero-Ugarte, M., Tapia, J., Molina, L., Lazarte-Rivera, A., Pacheco-Salazar, D. G., & Esparza, M. (2018). Bioelectrogenesis with microbial fuel cells (MFCs) using the microalga Chlorella vulgaris and bacterial communities. *Electronic Journal of Biotechnology, 31,* 34–43.

Ieropoulos, I., Greenman, J., & Melhuish, C. (2012). Urine utilisation by microbial fuel cells; energy fuel for the future. *Physical Chemistry Chemical Physics, 14*(1), 94–98.

Jansen, M. A. K., Mattoo, A. K., Malkin, S., & Edelman, M. (1993). Direct demonstration of binding-site competition between photosystem II inhibitors at the QB niche of the D1 protein. *Pesticide Biochemistry and Physiology, 46*(1), 78–83.

Jiang, Y., Zhang, W., Wang, J., Chen, Y., Shen, S., & Liu, T. (2013). Utilization of simulated flue gas for cultivation of *Scenedesmus dimorphus*. *Bioresource Technology, 128*, 359–364.

Juang, D. F., Yang, P. C., & Kuo, T. H. (2012). Effects of flow rate and chemical oxygen demand removal characteristics on power generation performance of microbial fuel cells. *International Journal of Environmental Science and Technology, 9*(2), 267–280.

Kato, M., Zhang, J. Z., Paul, N., & Reisner, E. (2014). Protein film photoelectrochemistry of the water oxidation enzyme photosystem II. *Chemical Society Reviews, 43*(18), 6485–6497.

Kawagoshi, Y., Hino, N., Fujimoto, A., Nakao, M., Fujita, Y., Sugimura, S., & Furukawa, K. (2005). Effect of inoculum conditioning on hydrogen fermentation and pH effect on bacterial community relevant to hydrogen production. *Journal of Bioscience and Bioengineering, 100*(5), 524–530.

Khalfbadam, H. M., Ginige, M. P., Sarukkalige, R., Kayaalp, A. S., & Cheng, K. Y. (2016). Bioelectrochemical system as an oxidising filter for soluble and particulate organic matter removal from municipal wastewater. *Chemical Engineering Journal, 296*, 225–233.

Khandelwal, A., Vijay, A., Dixit, A., & Chhabra, M. (2018). Microbial fuel cell powered by lipid extracted algae: A promising system for algal lipids and power generation. *Bioresource Technology, 247*, 520–527. https://doi.org/10.1016/j.biortech.2017.09.119

Kim, L. H., Kim, Y. J., Hong, H., Yang, D., Han, M., Yoo, G., Song, H. W., Chae, Y., Pyun, J., & Grossman, A. R. (2016). Patterned nanowire electrode array for direct extraction of photosynthetic electrons from multiple living algal cells. *Advanced Functional Materials, 26*(42), 7679–7689.

Kim, Y. J., Yun, J., Kim, S. Il, Hong, H., Park, J.-H., Pyun, J.-C., & Ryu, W. (2018). Scalable long-term extraction of photosynthetic electrons by simple sandwiching of nanoelectrode array with densely-packed algal cell film. *Biosensors and Bioelectronics, 117*, 15–22.

Kless, H., Oren-Shamir, M., Malkin, S., McIntosh, L., & Edelman, M. (1994). The DE region of the D1 protein is involved in multiple quinone and herbicide interactions in photosystem II. *Biochemistry, 33*(34), 10501–10507.

Kokabian, B., & Gude, V. G. (2013). Photosynthetic microbial desalination cells (PMDCs) for clean energy, water and biomass production. *Environmental Science: Processes & Impacts, 15*(12), 2178–2185.

Kondaveeti, S., Kakarla, R., & Min, B. (2017). Physicochemical parameters governing microbial fuel cell performance. In: Das, D. (eds), *Microbial Fuel Cell: A Bioelectrochemical System that Converts Waste to Watts* (pp. 189–208). https://doi.org/10.1007/978-3-319-66793-5_10

Kornet, J. G., Griffioen, H., & Schurer, K. (1992). An improved sensor for the measurement of chlorophyll fluorescence. *Measurement Science and Technology, 3*(2), 196.

Kremer, T. A., LaSarre, B., Posto, A. L., & McKinlay, J. B. (2015). N_2 gas is an effective fertilizer for bioethanol production by *Zymomonas mobilis*. *Proceedings of the National Academy of Sciences, 112*(7), 2222–2226.

Lee, D.-J., Chang, J.-S., & Lai, J.-Y. (2015). Microalgae–microbial fuel cell: A mini review. *Bioresource Technology, 198*, 891–895.

Lee, H., & Choi, S. (2015). A micro-sized bio-solar cell for self-sustaining power generation. *Lab on a Chip, 15*(2), 391–398.

Li, L. H., Sun, Y. M., Yuan, Z. H., Kong, X. Y., & Li, Y. (2013). Effect of temperature change on power generation of microbial fuel cell. *Environmental Technology*, *34*(13–14), 1929–1934.

Lin, C.-C., Wei, C.-H., Chen, C.-I., Shieh, C.-J., & Liu, Y.-C. (2013). Characteristics of the photosynthesis microbial fuel cell with a *Spirulina platensis* biofilm. *Bioresource Technology*, *135*, 640–643.

Liu, W., & Cheng, S. (2014). Microbial fuel cells for energy production from wastewaters: The way toward practical application. *Journal of Zhejiang University SCIENCE A*, *15*(11), 841–861.

Liu, Y., Climent, V., Berna, A., & Feliu, J. M. (2011). Effect of temperature on the catalytic ability of electrochemically active biofilm as anode catalyst in microbial fuel cells. *Electroanalysis*, *23*(2), 387–394.

Logan, B. E., Wallack, M. J., Kim, K. Y., He, W., Feng, Y., & Saikaly, P. E. (2015). Assessment of microbial fuel cell configurations and power densities. *Environmental Science and Technology Letters*, *2*, 206–214. https://doi.org/10.1021/acs.estlett.5b00180.

Mahmoud, R. H., Abdo, S. M., Samhan, F. A., Ibrahim, M. K., Ali, G. H., & Hassan, R. Y. A. (2020). Biosensing of algal-photosynthetic productivity using nanostructured bioelectrochemical systems. *Journal of Chemical Technology & Biotechnology*, *95*(4), 1028–1037.

Mahmoud, R. H., Samhan, F. A., Ali, G. H., Ibrahim, M. K., & Hassan, R. Y. A. (2018). Assisting the biofilm formation of exoelectrogens using nanostructured microbial fuel cells. *Journal of Electroanalytical Chemistry*, *824*(July), 128–135. https://doi.org/10.1016/j.jelechem.2018.07.045

Markou, G., Angelidaki, I., Nerantzis, E., & Georgakakis, D. (2013). Bioethanol production by carbohydrate-enriched biomass of *Arthrospira* (Spirulina) *platensis*. *Energies*, *6*(8), 3937–3950.

McCormick, A. J., Bombelli, P., Bradley, R. W., Thorne, R., Wenzel, T., & Howe, C. J. (2015). Biophotovoltaics: Oxygenic photosynthetic organisms in the world of bioelectrochemical systems. *Energy & Environmental Science*, *8*(4), 1092–1109.

Mendez, L., Mahdy, A., Ballesteros, M., & González-Fernández, C. (2014). Methane production of thermally pretreated *Chlorella vulgaris* and *Scenedesmus* sp. biomass at increasing biomass loads. *Applied Energy*, *129*, 238–242.

Michie, I. S., Kim, J. R., Dinsdale, R. M., Guwy, A. J., & Premier, G. C. (2011). The influence of psychrophilic and mesophilic start-up temperature on microbial fuel cell system performance. *Energy & Environmental Science*, *4*(3), 1011–1019.

Mingazzini, M., Saenz, M. E., Albergoni, F. G., & Marre, M. T. (1997). Algal photosynthesis measurements in toxicity testing. *Fresenius Environmental Bulletin*, *6*(5), 308–313.

Mohamed, S. N., Hiraman, P. A., Muthukumar, K., & Jayabalan, T. (2020). Bioelectricity production from kitchen wastewater using microbial fuel cell with photosynthetic algal cathode. *Bioresource Technology*, *295*, 122226.

Moriuchi, T., Morishima, K., & Furukawa, Y. (2008). Improved power capability with pyrolyzed carbon electrodes in micro direct photosynthetic/metabolic bio-fuel cell. *International Journal of Precision Engineering and Manufacturing*, *9*(2), 23–27.

Naessens, M., Leclerc, J. C., & Tran-Minh, C. (2000). Fiber optic biosensor using *Chlorella vulgaris* for determination of toxic compounds. *Ecotoxicology and Environmental Safety*, *46*(2), 181–185. https://doi.org/10.1006/eesa.1999.1904

Naessens, M., & Tran-Minh, C. (1998). Whole-cell biosensor for determination of volatile organic compounds in the form of aerosols. *Analytica Chimica Acta*, *364*(1–3), 153–158.

Nastro, R. A. (2014). Microbial fuel cells in waste treatment: Recent advances. *International Journal of Performability Engineering*, *10*, 367–376. doi: 10.23940/ ijpe.14.4.p367.mag

Nastro, R. A., Jannelli, N., Minutillo, M., Guida, M., Trifuoggi, M., Andreassi, L., Facci, A. L., Krastev, V. K., & Falcucci, G. (2017). Performance evaluation of microbial fuel cells fed by solid organic waste: Parametric comparison between three generations. *Energy Procedia*, *105*, 1102–1108. https://doi.org/10.1016/ j.egypro.2017.03.472

Nayak, J. K., & Ghosh, U. K. (2020). Microalgae cultivation for pre-treatment of pharmaceutical wastewater associated with microbial fuel cell and biomass feed stock production. In: V. Naddeo, M. Balakrishnan, & K.-H. Choo (eds.), *Frontiers in water-energy-nexus—Nature-based solutions, advanced technologies and best practices for environmental sustainability* (pp. 383–387). Springer.

Neethu, B., Pradhan, H., Sarkar, P., & Ghangrekar, M. M. (2019). Application of ion exchange membranes in enhancing algal production alongside desalination of saline water in microbial fuel cell. *MRS Advances*, *4*(19), 1077–1085.

Nguyen, H. T. H., Kakarla, R., & Min, B. (2017). Algae cathode microbial fuel cells for electricity generation and nutrient removal from landfill leachate wastewater. *International Journal of Hydrogen Energy*, *42*(49), 29433–29442. https://doi.org/ 10.1016/j.ijhydene.2017.10.011

Nishio, K., Hashimoto, K., & Watanabe, K. (2013). Light/electricity conversion by defined cocultures of Chlamydomonas and Geobacter. *Journal of Bioscience and Bioengineering*, *115*(4), 412–417. https://doi.org/10.1016/j.jbiosc.2012.10.015

Odhong, C., Wilkes, A., van Dijk, S., Vorlaufer, M., Ndonga, S., Sing'ora, B., & Kenyanito, L. (2019). Financing large-scale mitigation by smallholder farmers: What roles for public climate finance? *Frontiers in Sustainable Food Systems*, *3*, 3.

Oliveira, V. B., Simões, M., Melo, L. F., & Pinto, A. (2013). Overview on the developments of microbial fuel cells. *Biochemical Engineering Journal*, *73*, 53–64.

Pandard, P., & Rawson, D. M. (1993). An amperometric algal biosensor for herbicide detection employing a carbon cathode oxygen electrode. *Environmental Toxicology and Water Quality*, *8*(3), 323–333.

Pandard, P., Vasseur, P., & Rawson, D. M. (1993). Comparison of two types of sensors using eukaryotic algae to monitor pollution of aquatic systems. *Water Research*, *27*(3), 427–431.

Patil, P. D., Gude, V. G., Mannarswamy, A., Cooke, P., Nirmalakhandan, N., Lammers, P., & Deng, S. (2012). Comparison of direct transesterification of algal biomass under supercritical methanol and microwave irradiation conditions. *Fuel*, *97*, 822–831.

Patil, S. A., Harnisch, F., Koch, C., Hübschmann, T., Fetzer, I., Carmona-Martínez, A. A., Müller, S., & Schröder, U. (2011). Electroactive mixed culture derived biofilms in microbial bioelectrochemical systems: the role of pH on biofilm formation, performance and composition. *Bioresource Technology*, *102*(20), 9683–9690.

Peña-Vázquez, E., Pérez-Conde, C., Costas, E., & Moreno-Bondi, M. C. (2010). Development of a microalgal PAM test method for Cu (II) in waters: Comparison of using spectrofluorometry. *Ecotoxicology*, *19*(6), 1059–1065.

Powell, E. E., Mapiour, M. L., Evitts, R. W., & Hill, G. A. (2009). Growth kinetics of *Chlorella vulgaris* and its use as a cathodic half cell. *Bioresource Technology*, *100*(1), 269–274.

Rashid, N., Cui, Y.-F., Saif Ur Rehman, M., & Han, J.-I. (2013). Enhanced electricity generation by using algae biomass and activated sludge in microbial fuel cell. *The*

Science of the Total Environment, 456–457, 91–94. https://doi.org/10.1016/j.scitotenv.2013.03.067

Ren, H., Lee, H.-S., & Chae, J. (2012). Miniaturizing microbial fuel cells for potential portable power sources: Promises and challenges. *Microfluidics and Nanofluidics, 13*(3), 353–381.

Rioboo, C., González, O., Herrero, C., & Cid, A. (2002). Physiological response of freshwater microalga (*Chlorella vulgaris*) to triazine and phenylurea herbicides. *Aquatic Toxicology, 59*(3), 225–235. https://doi.org/10.1016/S0166-445X(01)00255-7

Rismani-Yazdi, H., Carver, S. M., Christy, A. D., & Tuovinen, O. H. (2008). Cathodic limitations in microbial fuel cells: An overview. *Journal of Power Sources, 180*(2), 683–694. Elsevier. https://doi.org/10.1016/j.jpowsour.2008.02.074

Rodrigo, M. A., Cañizares, P., García, H., Linares, J. J., & Lobato, J. (2009). Study of the acclimation stage and of the effect of the biodegradability on the performance of a microbial fuel cell. *Bioresource Technology, 100*(20), 4704–4710.

Rosenbaum, M., He, Z., & Angenent, L. T. (2010). Light energy to bioelectricity: Photosynthetic microbial fuel cells. *Current Opinion in Biotechnology, 21*(3), 259–264. https://doi.org/10.1016/j.copbio.2010.03.010

Rossi, D. M., da Costa, J. B., de Souza, E. A., Peralba, M. do C. R., & Ayub, M. A. Z. (2012). Bioconversion of residual glycerol from biodiesel synthesis into 1, 3-propanediol and ethanol by isolated bacteria from environmental consortia. *Renewable Energy, 39*(1), 223–227.

Rouillona, R., Tocabens, M., & Carpentier, R. (1999). A photoelectrochemical cell for detecting pollutant-induced effects on the activity of immobilized cyanobacterium Synechococcus sp. PCC 7942. *Enzyme and Microbial Technology, 25*(3–5), 230–235.

Ryu, W., Bai, S.-J., Park, J. S., Huang, Z., Moseley, J., Fabian, T., Fasching, R. J., Grossman, A. R., & Prinz, F. B. (2010). Direct extraction of photosynthetic electrons from single algal cells by nanoprobing system. *Nano Letters, 10*(4), 1137–1143.

Safi, C., Zebib, B., Merah, O., Pontalier, P.-Y., & Vaca-Garcia, C. (2014). Morphology, composition, production, processing and applications of *Chlorella vulgaris*: A review. *Renewable and Sustainable Energy Reviews, 35,* 265–278.

Saifuddin, N., & Priatharsini, P. (2016). Developments in bio-hydrogen production from algae: A review. *Research Journal of Applied Sciences, Engineering and Technology, 12*(9), 968–982. https://doi.org/10.19026/rjaset.12.2815

Salar-García, M. J., Gajda, I., Ortiz-Martínez, V. M., Greenman, J., Hanczyc, M. M., de Los Ríos, A. P., & Ieropoulos, I. A. (2016). Microalgae as substrate in low cost terracotta-based microbial fuel cells: Novel application of the catholyte produced. *Bioresource Technology, 209,* 380–385.

Saratale, R. G., Kuppam, C., Mudhoo, A., Saratale, G. D., Periyasamy, S., Zhen, G., Koók, L., Bakonyi, P., Nemestóthy, N., & Kumar, G. (2017a). Bioelectrochemical systems using microalgae – A concise research update. *Chemosphere, 177,* 35–43.

Saratale, R. G., Kuppam, C., Mudhoo, A., Saratale, G. D., Periyasamy, S., Zhen, G., Koók, L., Bakonyi, P., Nemestóthy, N., & Kumar, G. (2017b). Bioelectrochemical systems using microalgae – A concise research update. *Chemosphere, 177,* 35–43. https://doi.org/10.1016/j.chemosphere.2017.02.132

Sayegh, A., Longatte, G., Buriez, O., Wollman, F.-A., Guille-Collignon, M., Labbé, E., Delacotte, J., & Lemaître, F. (2019). Diverting photosynthetic electrons from suspensions of Chlamydomonas reinhardtii algae – New insights using an electrochemical well device. *Electrochimica Acta, 304,* 465–473.

Schenk, P. M., Thomas-Hall, S. R., Stephens, E., Marx, U. C., Mussgnug, J. H., Posten, C., Kruse, O., & Hankamer, B. (2008). Second generation biofuels: High-efficiency microalgae for biodiesel production. *BioEnergy Research, 1*(1), 20–43. https://doi.org/10.1007/s12155-008-9008-8

Scognamiglio, V., Pezzotti, I., Pezzotti, G., Cano, J., Manfredonia, I., Buonasera, K., Arduini, F., Moscone, D., Palleschi, G., & Giardi, M. T. (2012). Towards an integrated biosensor array for simultaneous and rapid multi-analysis of endocrine disrupting chemicals. *Analytica Chimica Acta, 751*, 161–170.

Scognamiglio, V., Pezzotti, I., Pezzotti, G., Cano, J., Manfredonia, I., Buonasera, K., Rodio, G., & Giardi, M. T. (2013). A new embedded biosensor platform based on micro-electrodes array (MEA) technology. *Sensors and Actuators B: Chemical, 176*, 275–283.

Sekar, N., & Ramasamy, R. P. (2015). Recent advances in photosynthetic energy conversion. *Journal of Photochemistry and Photobiology C: Photochemistry Reviews, 22*, 19–33.

Sekar, N., Umasankar, Y., & Ramasamy, R. P. (2014). Photocurrent generation by immobilized cyanobacteria via direct electron transport in photo-bioelectrochemical cells. *Physical Chemistry Chemical Physics, 16*, 7862–7871. https://doi.org/10.1039/c4cp00494a.

Shaishav, S., Singh, R. N., & Satyendra, T. (2013). Biohydrogen from algae: Fuel of the future. *International Research Journal of Environmental Sciences, 2*(4), 44–47.

Shitanda, I., Takada, K., Sakai, Y., & Tatsuma, T. (2005a). Amperometric biosensing systems based on motility and gravitaxis of flagellate algae for aquatic risk assessment. *Analytical Chemistry, 77*(20), 6715–6718.

Shitanda, I., Takada, K., Sakai, Y., & Tatsuma, T. (2005b). Compact amperometric algal biosensors for the evaluation of water toxicity. *Analytica Chimica Acta, 530*(2), 191–197.

Shukla, M., & Kumar, S. (2018). Algal growth in photosynthetic algal microbial fuel cell and its subsequent utilization for biofuels. *Renewable and Sustainable Energy Reviews, 82*(June 2017), 402–414. https://doi.org/10.1016/j.rser.2017.09.067

Simkin, A. J., López-Calcagno, P. E., & Raines, C. A. (2019). Feeding the world: improving photosynthetic efficiency for sustainable crop production. *Journal of Experimental Botany, 70*(4), 1119–1140.

Song, X., Wang, W., Cao, X., Wang, Y., Zou, L., Ge, X., Zhao, Y., Si, Z., & Wang, Y. (2020). Chlorella vulgaris on the cathode promoted the performance of sediment microbial fuel cells for electrogenesis and pollutant removal. *Science of the Total Environment, 728*, 138011.

Strik, D. P. B. T. B. B. T. B., Timmers, R. A., Helder, M., Steinbusch, K. J. J. J., Hamelers, H. V. M. M., & Buisman, C. J. N. N. (2011). Microbial solar cells: Applying photosynthetic and electrochemically active organisms. *Trends in Biotechnology, 29*(1), 41–49. https://doi.org/10.1016/j.tibtech.2010.10.001

Sulfahri Amin, M., Sumitro, S. B., & Saptasari, M. (2016). Bioethanol production from algae *Spirogyra hyalina* using *Zymomonas mobilis*. *Biofuels, 7*(6), 621–626.

Sulfahri, M. S., Sunarto, E., Irvansyah, M. Y., Utami, R. S., & Mangkoedihardjo, S. (2011). Ethanol production from algae Spirogyra with fermentation by *Zymomonas mobilis* and *Saccharomyces cerevisiae*. *Journal of Basic and Applied Scientific Research, 1*(7), 589–593.

Tang, Y. L., He, Y. T., Yu, P. F., Sun, H., & Fu, J. X. (2012). Effect of temperature on electricity generation of single-chamber microbial fuel cells with proton exchange membrane. *Advanced Materials Research, 393*, 1169–1172.

Tanisho, S., Kamiya, N., & Wakao, N. (1989). Microbial fuel cell using Enterobacter aerogenes. *Bioelectrochemistry and Bioenergetics*, *21*(1), 25–32. https://doi.org/10.1016/0302-4598(89)87003-5

Tel-Vered, R., & Willner, I. (2014). Photo-bioelectrochemical cells for energy conversion, sensing, and optoelectronic applications. *ChemElectroChem*, *1*(11), 1778–1797.

Turbak, S. C., Olson, S. B., & McFeters, G. A. (1986). Comparison of algal assay systems for detecting waterborne herbicides and metals. *Water Research*, *20*(1), 91–96.

Turemis, M., Silletti, S., Pezzotti, G., Sanchís, J., Farré, M., & Giardi, M. T. (2018). Optical biosensor based on the microalga-paramecium symbiosis for improved marine monitoring. *Sensors and Actuators B: Chemical*, *270*, 424–432.

Ucar, D., Zhang, Y., & Angelidaki, I. (2017). An overview of electron acceptors in microbial fuel cells. *Frontiers in Microbiology*, *8*, 643.

Ullah, K., Ahmad, M., Sharma, V. K., Lu, P., Harvey, A., Zafar, M., Sultana, S., & Anyanwu, C. N. (2014). Algal biomass as a global source of transport fuels: Overview and development perspectives. *Progress in Natural Science: Materials International*, *24*(4), 329–339.

Védrine, C., Fabiano, S., & Tran-Minh, C. (2003). Amperometric tyrosinase based biosensor using an electrogenerated polythiophene film as an entrapment support. *Talanta*, *59*(3), 535–544.

Védrine, C., Leclerc, J.-C., Durrieu, C., & Tran-Minh, C. (2003). Optical whole-cell biosensor using *Chlorella vulgaris* designed for monitoring herbicides. *Biosensors and Bioelectronics*, *18*(4), 457–463.

Velasquez-Orta, S. B., Curtis, T. P., & Logan, B. E. (2009). Energy from algae using microbial fuel cells. *Biotechnology and Bioengineering*, *103*(6), 1068–1076. https://doi.org/10.1002/bit.22346

Venkata Mohan, S., Srikanth, S., Chiranjeevi, P., Arora, S., & Chandra, R. (2014). Algal biocathode for in situ terminal electron acceptor (TEA) production: Synergetic association of bacteria-microalgae metabolism for the functioning of biofuel cell. *Bioresource Technology*, *166*(December), 566–574. https://doi.org/10.1016/j.biortech.2014.05.081

Wang, H., Qian, F., & Li, Y. (2014). Solar-assisted microbial fuel Cells for bioelectricity and chemical fuel generation. *Nano Energy*, *8*, 264–273. https://doi.org/10.1016/j.nanoen.2014.06.004

Wang, X., Feng, Y., Liu, J., Lee, H., Li, C., Li, N., & Ren, N. (2010). Sequestration of CO_2 discharged from anode by algal cathode in microbial carbon capture cells (MCCs). *Biosensors and Bioelectronics*, *25*(12), 2639–2643. https://doi.org/10.1016/j.bios.2010.04.036

Wang, Z., Deng, H., Chen, L., Xiao, Y., & Zhao, F. (2013). In situ measurements of dissolved oxygen, pH and redox potential of biocathode microenvironments using microelectrodes. *Bioresource Technology*, *132*, 387–390.

Wong, S. L., & Beaver, J. L. (1980). Algal bioassays to determine toxicity of metal mixtures. *Hydrobiologia*, *74*(3), 199–208.

Wu, X., Song, T., Zhu, X., Wei, P., & Zhou, C. C. (2013). Construction and operation of microbial fuel cell with *Chlorella vulgaris* biocathode for electricity generation. *Applied Biochemistry and Biotechnology*, *171*(8), 2082–2092. http://www.ncbi.nlm.nih.gov/pubmed/24404595

Wu, Y., Guan, K., Wang, Z., Xu, B., & Zhao, F. (2013). Isolation, identification and characterization of an electrogenic microalgae strain. *PLoS ONE*, *8*(9), 1–7. https://doi.org/10.1371/journal.pone.0073442

Xiao, L., & He, Z. (2014). Applications and perspectives of phototrophic microorganisms for electricity generation from organic compounds in microbial fuel cells. *Renewable and Sustainable Energy Reviews*, 37, 550–559. https://doi.org/10.1016/j.rser.2014.05.066

Xu, C., Poon, K., Choi, M. M. F., & Wang, R. (2015). Using live algae at the anode of a microbial fuel cell to generate electricity. *Environmental Science and Pollution Research*, 22(20), 15621–15635. https://doi.org/10.1007/s11356-015-4744-8

Yadavalli, R., Ratnapuram, H., Motamarry, S., Reddy, C. N., Ashokkumar, V., & Kuppam, C. (2020). Simultaneous production of flavonoids and lipids from Chlorella vulgaris and Chlorella pyrenoidosa. *Biomass Conversion and Biorefinery*, 12, 683–691. https://doi.org/10.1007/s13399-020-01044-x.

Yadavalli, R., Ratnapuram, H., Peasari, J. R., Reddy, C. N., Ashokkumar, V., & Kuppam, C. (2021). Simultaneous production of astaxanthin and lipids from *Chlorella sorokiniana* in the presence of reactive oxygen species: A biorefinery approach. *Biomass Conversion and Biorefinery*, 12, 1–9. https://doi.org/10.1007/s13399-021-01276-5.

Yang, L.-H., Zhu, T.-T., Cai, W.-W., Haider, M. R., Wang, H.-C., Cheng, H.-Y., & Wang, A.-J. (2018). Micro-oxygen bioanode: An efficient strategy for enhancement of phenol degradation and current generation in mix-cultured MFCs. *Bioresource Technology*, 268, 176–182.

Yang, Z., Nie, C., Hou, Q., Zhang, L., Zhang, S., Yu, Z., & Pei, H. (2019). Coupling a photosynthetic microbial fuel cell (PMFC) with photobioreactors (PBRs) for pollutant removal and bioenergy recovery from anaerobically digested effluent. *Chemical Engineering Journal*, 359, 402–408.

Yang, Z., Pei, H., Hou, Q., Jiang, L., Zhang, L., & Nie, C. (2018). Algal biofilm-assisted microbial fuel cell to enhance domestic wastewater treatment: Nutrient, organics removal and bioenergy production. *Chemical Engineering Journal*, 332, 277–285.

Yang, Z., Guo, R., Xu, X., Fan, X., & Luo, S. (2011). Fermentative hydrogen production from lipid-extracted microalgal biomass residues. *Applied Energy*, 88(10), 3468–3472.

Yeruva, D. K., Velvizhi, G., & Mohan, S. V. (2016). Coupling of aerobic/anoxic and bioelectrogenic processes for treatment of pharmaceutical wastewater associated with bioelectricity generation. *Renewable Energy*, 98, 171–177.

Zhang, T., Zeng, Y., Chen, S., Ai, X., & Yang, H. (2007). Improved performances of E. coli-catalyzed microbial fuel cells with composite graphite/PTFE anodes. *Electrochemistry Communications*, 9(3), 349–353. https://doi.org/10.1016/j.elecom.2006.09.025

Zhang, X., Wang, Q., Tang, C., Wang, H., Liang, P., Huang, X., & Zhang, Q. (2020). High-power microbial fuel cells based on a carbon–carbon composite air cathode. *Small*, 16(15), 1905240.

Zhang, Y., Noori, J. S., & Angelidaki, I. (2011). Simultaneous organic carbon, nutrients removal and energy production in a photomicrobial fuel cell (PFC). *Energy & Environmental Science*, 4(10), 4340–4346.

Zhang, Y., Zhao, Y., & Zhou, M. (2019). A photosynthetic algal microbial fuel cell for treating swine wastewater. *Environmental Science and Pollution Research*, 26(6), 6182–6190.

Zhou, M., He, H., Jin, T., & Wang, H. (2012). Power generation enhancement in novel microbial carbon capture cells with immobilized *Chlorella vulgaris*. *Journal of Power Sources*, 214, 216–219.

5 An Overview on Low-Cost Anode Materials and Their Modifications in Microbial Fuel Cells (MFCs) towards Enhancement in Performance

Sanath Kondaveeti, Aarti Bisht, and Raviteja Pagolu
Department of Chemical Engineering, Konkuk University, Seoul, Republic of Korea

C. Nagendranatha Reddy
Department of Biotechnology, Chaitanya Bharathi Institute of Technology (Autonomous), Gandipet, Hyderabad, Telangana State, India

K. Chandrasekhar
School of Civil and Environmental Engineering, Yonsei University, Seoul, Republic of Korea

Jung-Kul Lee
Department of Chemical Engineering, Konkuk University, Seoul, Republic of Korea

CONTENTS

DOI: 10.1201/9781003225430-5

5.1   INTRODUCTION

An increase in pollution and energy shortage have deleterious effects on the overall environment thereby impacting human health and various ecosystems (Mal et al., 2021; Raj et al., 2021a). There is a growing concern in using substitute energy that is renewable and self-sustainable without any negative impact on the environment (Chandrasekhar et al., 2020a; Chandrasekhar et al., 2021b; Park et al., 2021; Raj et al., 2021b). Also, rapid urbanization and industrialization, along with rising energy consumption and a growing global population, have a considerable influence on pollution in the environment (Kondaveeti et al., 2014a; Kondaveeti et al., 2014b). In this regard, microbial fuel cells (MFCs) appear to be a viable solution for addressing such issues (Chandrasekhar et al., 2021c; Chandrasekhar et al., 2021d; Gambino et al., 2021). In general, MFCs generate energy by converting organic matter into power via bioelectrogenic microbes through their bacterial metabolism. MFC includes working and counter electrodes divided by a proton permeable membrane in MFCs (PEM). Anodic bioconsortia oxidizes the given substrate (organic) to release H^+ (protons) and e^- (electrons). The circuit connected externally between the working and counter electrodes transports electrons, whereas protons diffuse via PEM to the cathode chamber (Mohanakrishna et al., 2018). On the cathode, they combine with oxygen from the air to create water (Chandrasekhar et al., 2021a; Chandrasekhar et al., 2020b; Enamala et al., 2020). MFCs have several benefits over previous conventional techniques regarding power generation and organic removal, including reduced activated sludge formation; moreover, the possibility to work without any external aeration energy, easy operation under normal conditions, and environmental friendliness (Mohan & Chandrasekhar, 2011a; Mohan & Chandrasekhar, 2011b).

Inadequate energy generation over conventional fuel cells and high cost make the MFCs currently unsuitable for commercial and/or industrial use. In MFCs, the selection of suitable material along with design and fabrication still pose a significant difficulty. Conductive metals, low-cost carbon-based materials, and electrically conducting polymers are investigated as potential anode electrode materials for MFCs (Sarathi & Nahm, 2013). However, traditional carbon materials like carbon paper, graphite rods, carbon brushes, and carbon cloth are effectively tested. Conductive metals like gold, silver, copper, and conducting polymers like polypyrrole and polyaniline are used (Wei et al., 2011). However, in a few cases, these materials are limited with chemical instability, inadequate porosity and surface area, unstable mechanical property, biological compatibility, corrosiveness, and elevated budget. As a result, high-quality material is necessary to overcome these disadvantages and produce a stable anode (Cai et al., 2020).

This chapter gives a summary of anode material types as well as common modification techniques for increasing anode functioning. In addition, different natural materials and modifying techniques were widely examined to meet the existing problems of anode electrodes in MFCs, like stability and cost.

5.2 IMPORTANT FEATURES OF ANODE MATERIAL IN MFCs

In order to get suitable performance in terms of electrochemical efficacy, electron transfer rate, and bioelectrogens adherence, researchers must still choose the efficient anode electrode material for MFCs. Therefore, few critical properties that must be noted in a perfect anode electrode are discussed accordingly.

Due to the transfer of electrons from anode to the cathode through an external circuit, conductivity is an essential feature of anode materials (Chandrasekhar et al., 2021c; Chandrasekhar et al., 2021d; Sun et al., 2011). As a result, the anode material is in charge of allowing electrons to flow freely and boosting their transfer rate. In general, highly conductive materials aid in lowering the bulk resistance of electrolytes and increasing electron transport. Also, to increase electron transport, the interfacial resistance among the biogenic electrode and substrate in the electrolyte must be minimal. Therefore, before constructing the anode electrodes for MFCs, the electrical conductivity of materials is generally investigated. Based on the earlier studies, electrode surface area has a significant impact on energy production in MFCs (Chandrasekhar & Ahn, 2017; Deval et al., 2017; Gambino et al., 2021). The resistance offered by the electrode is unambiguously related to the ohmic losses of the MFC, therefore increasing its surface area is the most straightforward approach to minimize resistance power. Furthermore, a higher surface area provides a more functional area for the growth of bioelectrogens and improves the electrode kinetics efficiency. Bioelectrogenic microbes like *Geobacter, Shewanella,* and various other species, were effectively and dynamically bound on working electrode surfaces, guaranteeing adequate and direct e⁻ transmission (Sun et al., 2011). Several metabolic processes related to electron transfer of bioelectrogenic active microbes take place on the anode surface. Therefore, the surface area of an anode has a significant impact on MFC performance (Kamedulski et al., 2019). The other studies noted that conventional carbon-built materials like graphene and its products have a larger surface area than typical carbon compounds like carbon paper or carbon cloth (Hindatu et al., 2017; Zhang et al., 2020).

The anode biocompatibility is also one of the critical MFC functions since it comes into close contact with bioelectrogenic microbes and their metabolic activity. Several materials, including Cu, Ag, and Au, are not biocompatible for use as the anode because of their corrosivity nature (Li et al., 2017a; Yaqoob et al., 2020b). Moreover, the toxicity of these metals can prevent bacterial development during MFC operation, thus resulting in overall lower energy output (Yamashita & Yokoyama, 2018). The long-term operation of MFC can lead to instability of anode, which can be due to variation in their characteristics, such as the decrease in biocompatibility, due to variation in mechanical and chemical instability (Yaqoob et al., 2020a). Also, the enduring interaction of the working electrode with the wastewater and bioelectrogens usually results in swelling. Because of this, the electrode's physical stability is completely disrupted. The significant reasons for swelling can be due to corrosion of anode electrode and variation in mechanical strength. Also, the anode surface should be coarse to separate water molecules and provide additional active sites for bacteria adhesion (Mehranian et al., 2010).

The cost and accessibility of anode materials are key considerations since they directly impact the total cost of MFCs. Au-, Ag-, and Pt-type metals are extremely valuable and not easy to obtain (Yamashita & Yokoyama, 2018). Therefore, the metal composites and alloy particles, and natural carbon-based materials (e.g. activated carbon) may be an appealing alternative to costly metals as anode electrode materials in MFCs.

5.3 MATERIALS USED AS ANODES IN MFCs

A wide range of functions like substrate oxidation by bioelectrogens, electron transfer rate and bioelectrogenic biofilm formation, can be altered by the type of fabrication and variety of materials for the anode. It is well known that the quality of the anode and its characteristics impact energy production directly (He et al., 2015; Sarathi & Nahm, 2013). The often-used electrodes materials in MFCs are noted to be conductive metals, low-cost carbon-based materials, and electrically conducting polymers (Figure 5.1). However, in the view of pilot-scale operation, these materials can be limited with high cost. Therefore, naturally derived carbon-based materials like graphene, synthesized from the wastes, are being explored (Purkait et al., 2017). Also, reworked materials like metal composites, ammonia pre-treated materials, nitrogen-doped materials, and heat and chemically cured materials are being used as anodes in MFCs (Cheng & Logan, 2007; Feng et al., 2010).

5.3.1 Use of Carbon-Built Materials in MFCs

Currently, the carbon materials are widely used as anodes in MFCs, due to their conductive nature with wide surface area and high compatibility with bioelectrogens. Moreover, carbon materials are well known for their, thermal and chemical stability. And are readily available, with the possibility in tuning in electron transmission rate. The familiar carbon materials in MFC anodes are carbon cloth, graphite rod, activated carbon, carbon paper, graphite felt, graphite, carbon nanotube, carbon brush, and reticulated vitreous carbon (Li et al., 2017b).

Carbon cloth is one of the most used anode materials in MFC. Even though it has a wider surface area than plain carbon, yet it has a substantially higher porosity owing to massive blank space prevalence. In general, carbon cloth can outperform the plain carbon paper in terms of flexibility, mechanical strength,

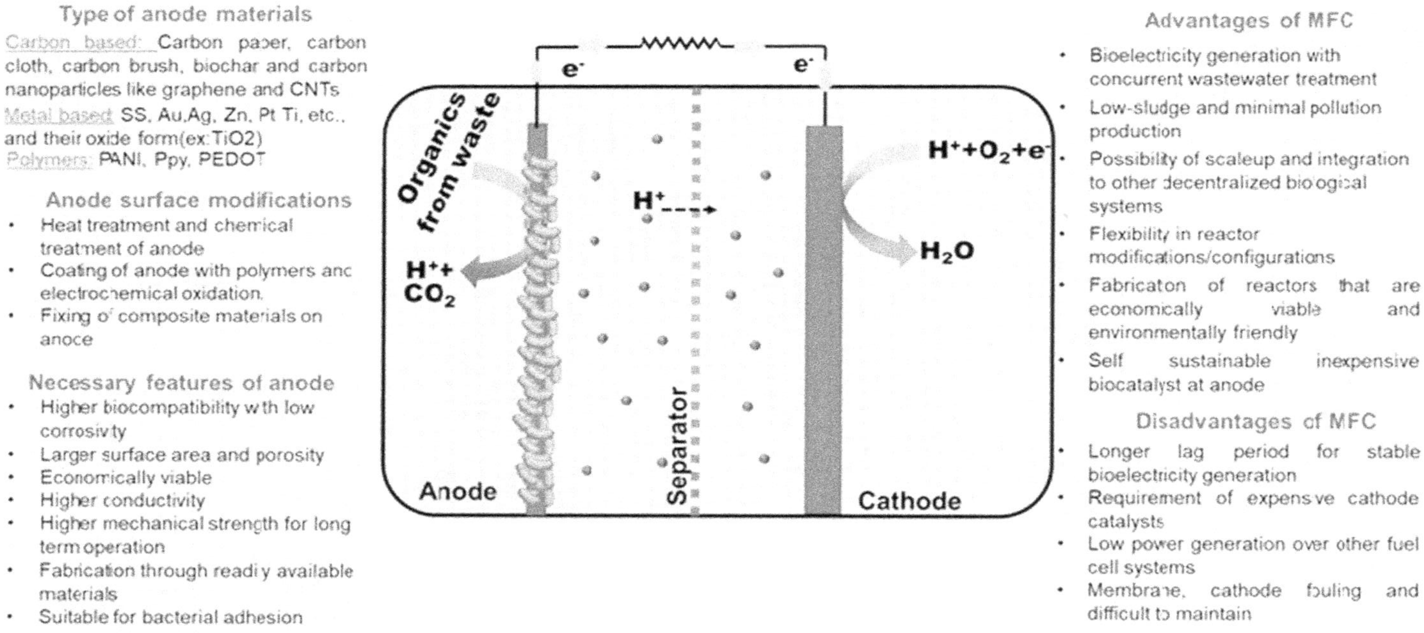
Type of anode materials
Carbon based: Carbon paper, carbon cloth, carbon brush, biochar and carbon nanoparticles like graphene and CNTs
Metal based: SS, Au,Ag, Zn, Pt Ti, etc., and their oxide form(ex:TiO2)
Polymers: PANI, Ppy, PEDOT
Anode surface modifications
• Heat treatment and chemical treatment of anode
• Coating of anode with polymers and electrochemical oxidation.
• Fixing of composite materials on anode
Necessary features of anode
• Higher biocompatibility with low corrosivity
• Larger surface area and porosity
• Economically viable
• Higher conductivity
• Higher mechanical strength for long term operation
• Fabrication through readily available materials
• Suitable for bacterial adhesion
e⁻
e⁻
Organics from waste
H⁺+ CO₂
H⁺
Separator
H⁺+O₂+e⁻
H₂O
Anode
Cathode
Advantages of MFC
• Bioelectricity generation with concurrent wastewater treatment
• Low-sludge and minimal pollution production
• Possibility of scaleup and integration to other decentralized biological systems
• Flexibility in reactor modifications/configurations
• Fabricaton of reactors that are economically viable and environmentally friendly
• Self sustainable inexpensive biocatalyst at anode
Disadvantages of MFC
• Longer lag period for stable bioelectricity generation
• Requirement of expensive cathode catalysts
• Low power generation over other fuel cell systems
• Membrane, cathode fouling and difficult to maintain

FIGURE 5.1 Graphic description of microbial fuel cell along with the anode material types and necessary features. Advantages and disadvantages of MFCs are also presented.

and electrical conductivity. High cost and chemical instability that lead to fouling during long-term operation are the major drawbacks of carbon cloth (Li et al., 2017b).

Even though the carbon papers offer a high porosity, but the high cost of the carbon paper limits its use as an anode in MFCs. In this regard, carbon fiber brush seems a noteworthy carbon product that can offer a superior surface area and ideal areal volume to the cathode chamber. The higher electrical brush is ensured by using the titanium wire as the current collector. Therefore, carbon brushes are commonly used anodes; further studies are aimed at lowering their cost (Zhang et al., 2017).

Carbon mesh is also available commercially at a cheap price; however, it has poor electrical conductivity and mechanical stability, which indicates inadequate stability. However, the higher surface area is obtained by the flexible nature of carbon mesh, thereby aiding in 3D electrodes fabrication. Wang et al. have studied the influence of carbon mesh in comparison to other carbon material viz. cloth and paper and reported the necessity of carbon mesh pre-treatment for achieving higher power generation in MFCs (Wang et al., 2011). Reticulated vitreous carbon is another carbon material that is moderately used as an anode in MFCs. Due to its larger porous construction, it can enable MFCs to achieve a higher power generation. However, its use as a stand-alone electrode can limit the MFC performance, due to its brittle nature (Yuan & Kim, 2008). Carbon felt is also a widely used anode electrode in MFCs. It is porous and has strong electrical conductivity. Moreover, it is relatively cheaper in comparison to previously mentioned carbon materials and possess good mechanical stability dependent on the material thickness (Kondaveeti et al., 2018).

Based on several biological and bioelectrochemical studies, granular activated carbon (GAC) is known to be a low-cost biostable material (Caizán-Juanarena et al., 2019). Due to empty areas in packing of GAC at the anode, it can minimize the electron transfer and thereby limiting the overall electrochemical activity (Caizán-Juanarena et al., 2019). Therefore, GAC has to be modified further to enhance the conductivity and to decrease porosity. An earlier study by Borsje et al. has demonstrated the influence of specific surface area (SSA) of GAC on MFC electricity generation (Borsje et al., 2016). The better performance of MFC is noted by using the high SSA over low SSA and suggested the porosity of the GAC also controls the MFCs' electrochemical performance. However, additional studies are required to point which critical aspects control the MFC performance that employs GAC. Li et al. have pursued a comparative study on MFCs' electrochemical performance by using GAC and CC as an anode material in a double-chamber cell. In this study, they have noted a 2.5-fold higher power generation with GAC over CC, along with an increase in Columbic efficiency. Also, this study has proven that the higher surface area materials like GAC enable/enhance the development of bioelectrogenic biofilm and thereby enhance the power generation with concurrent organic reduction (Li et al., 2010). However, the use of GAC at MFC anodes require an additional electrode support for electron harvesting.

Graphite is another carbon substance that is often utilized in MFCs as the anode. Graphite prospective qualities include a higher mechanical strength, along with a larger surface area (Chandrasekhar & Venkata Mohan, 2012; Chandrasekhar & Venkata Mohan, 2014a; Chandrasekhar & Venkata Mohan, 2014b; Kumar et al., 2012). Moreover, it is well known for its compatibility with bioelectrogens at the anode. However, the low conductivity of graphite often limits its commercial usage as an MFC anode. In general, graphite rods are graphite plates that are the most commonly used anode materials. Ter-Heijne et al. have made a comparative evaluation in MFC using the plain graphite plate and course graphite plate. In this study, they have noted that altering the graphite surface by enhancing the roughness has improved the power generation in MFC with simultaneous larger organic reduction (ter Heijne et al., 2008).

Over the last decade, graphene has emerged as a suitable carbon material for the MFC anode (Chandrasekhar, 2019); due to its promising traits like compatibility with bioelectrogens, higher electrical conductivity, stability, and mechanical strength. Several reviews have summarized the use of graphene in MFCs. The 3D graphene sponge fabricated by Chen et al. illustrated an enhancement in power generation in MFCs, in comparison to commercial graphite felt. They have suggested that the enhancement in power generation is attributed due to microporosity of graphene, which assists in the better attachment of bioelectrogens (Chen et al., 2014). However, a cubic meter of graphene sponge would cost around $2,000, thereby limiting its commercial usage. In another study, Mehdinia et al. had combined the reduced graphene oxide (rGO) along with tin oxide nanoparticles and noted a fivefold increase in power generation in comparison to control (only rGO). The increase in power generation noted in this study is due to the incorporation of metal oxide nanoparticles that can increase the conductivity of anodes and electrodes (Mehdinia et al., 2014). This in turn increases the electron transfer rate and overall power generation. Even though graphene and its metallic and nonmetallic modifications have exhibited an outstanding performance as anodes in MFCs, the high cost renders its commercial application and thus makes it impractical. In this regard, the formulation of graphene and its variants using natural wastes can be an alternative cost-ineffective strategy to satisfy the criteria as an anode in MFCs (Kong et al., 2020). The summarized power generation noted by using the various carbon-based and metal-based materials are presented in Table 5.1.

5.3.2 ANODES FROM NATURAL WASTE

The preparation of anodes using natural wastes and biomass has piqued interest, due to several advantages like preparation from recyclable resources along with their promising stability and durability. An intriguing application in this regard is the carbonization-based creation of layered corrugated carbon (LCC)-based anodes using cost-effective materials like recycled paper. Here, the increase in layer number has enhanced the power generation accordingly (3 layers = 200 A/m^2 and 6 layers = 390 A/m^2), thus by enabling the larger surface area for the growth

TABLE 5.1
Frequently Used Anode Materials in MFCs and Their Modifications

Anode type	Reactor type	Power density	Process highlights	Reference
Modification of carbon electrodes				
Carbon paper	Double chamber	600 mW/m^2	Testing the influence of electrode material in comparison to others	(Logan et al., 2007)
Carbon brush	Double chamber	750 mW/m^2	Increase in surface area increased power output	(Logan et al., 2007)
Carbon cloth	Single chamber	1070 mW/m^2	Limited surface area decreases energy output	(Logan et al., 2007)
GO-CNT on carbon paper	Double chamber	188 mW/m^2	Variation in power generation noted with change in bacteria type	(Hassan et al., 2012)
Graphene on carbon brush	Double chamber	1905 mW/m^2	Generation of microbial reduced graphene	(Yuan et al., 2012)
Graphene on carbon cloth	Double chamber	52.5 mW/m^2	Enhanced power generation of 2.7-fold noted using graphene in comparison to plain CC	(Liu et al., 2012)
Graphite rods	Double chamber	26 mW/m^2	Use of multi anode decreases organics	(Liu et al., 2004)
Graphite + PTFE	Single chamber	760 mW/m^2	PTFE can alter power output in MFC	(Zhang et al., 2007)
Graphite felt + PANI	Single chamber	2300 mW/m^2	2-fold increase in power output is noted by using PANI over control	(Zhao et al., 2010)
Graphite felt + AQDS + PEI	Double chamber	480 mW/m^2	Stable biofilm formation in comparison to control system	(Adachi et al., 2008)
Carbon NP + CC	Double chamber	705 mW/m^2	Variation of carbon NP can alter power output	(Yuan et al., 2009)
Carbon cloth treated with ammonia	Single chamber	1,970 mW/m^2	48% increase in power and 50% decrease in lag period noted with ammonia treatment	(Cheng & Logan, 2007)
Carbon cloth oxidized by electrochemical oxidation	Single chamber	939 mW/m^2	14.2% higher power output in comparison to control	(Liu et al., 2014)
Graphite felt + poly (aniline-co-o-aminophenol)	Double chamber	23.8 mW/m^2	35% higher power generation compared to control	(Li et al., 2011)
Textile fabric + CNT	Double chamber	1098 mW/m^2	68% higher power generation compared to carbon cloth	(Xie et al., 2011)

TABLE 5.1 (Continued)
Frequently Used Anode Materials in MFCs and Their Modifications

Anode type	Reactor type	Power density	Process highlights	Reference
Stainless steel mesh + CNT	Single chamber	147 mW/m²	49-fold higher comparison to plain SS	(Zhang et al., 2013)
MFC anodes using metals and metal oxide composites				
Carbon paper with gold	Double chamber	346 mW/m²	50% increase in power and 36% decrease in lag period noted with Au	(Guo et al., 2012)
Graphite with graphene and tin oxide	Double chamber	1,624 mW/m²	2.8 and 4.8-fold higher than RGO and control anodes, respectively	(Mehdinia et al., 2014)
Carbon cloth with Pd	Double chamber	605 mW/m²	Power generation and CE were increased by 14% and 31%, respectively	(Quan et al., 2015)
NiO+PANI+graphite felt	Double chamber	1,078 mW/m²	6.6-fold higher current compared to control	(Zhong et al., 2018)
Nano-Fe3C+ porous graphitic carbon	Single chamber	1,856 mW/m²	Electrogens are enriched by Fe3C, which speeds up electron transfer	(Hu et al., 2019)

of bioelectrogens. Moreover, the current densities reported in this study were higher over normal graphite felt (Chen et al., 2012b). Also, the electrode material derived from the natural waste can be economic since that is available at viable costs in comparison to the traditional ones. Moreover, these materials are proven to provide fundamental needed features to possessed for an anode. These fundamental features include the higher electron transfer rate and thereby achieve better electrokinetics.

Although the increase in the surface area enables to achieve a higher power generation in MFCs, the low internal resistance of the system controls the overall reaction. In relation to the MFCs' in-house resistance, Chen et al. have fabricated MFCs anode using the stem of *Hibiscus cannabinus* through carbonization (Chen et al., 2012a). The resulting product from the carbonization is used as anode material with stainless steel support. They have also reported a threefold upsurge in current generation using the carbonized material with SS support, in comparison to conventional graphite electrodes. In the other study, Karthikeyan et al. have analyzed the MFC performance by using the anode fabricated from carbonized corn stems and wild and king mushrooms. In this study, the carbonized corn stem as an anode electrode has exhibited an eightfold increase in bioelectrocatalytic current in comparison to conventional graphite electrodes (Karthikeyan et al., 2015). From this, it can be noted that the variation of initial material for carbonization can lead to variation in MFC performance due to

TABLE 5.2

An Overview of the Non-Renewable Low-Cost Biomass Materials Utilized for Construction of Anode in Microbial Fuel Cell

Source of material	Reactor type	Power generation	Process highlights	Reference
Coffee waste	Single chamber	3,927 mW/m^2	Fourfold higher power generation compared to activated carbon	(Hung et al., 2019)
Onion peel	Single chamber	742 mW/m^2	Onion derived AC having rich nitrogen can enhance anodic electron transfer	(Li et al., 2018)
Shells from chestnut	Single chamber	23.6 mW/m^2	2.3 times higher in comparison to plain carbon cloth	(Chen et al., 2018b)
Urchin-like structure from chestnut shells	Single chamber	759 mW/m^2	Higher performance is due to microscopic and macroscopic structure resulting from chestnut shells	(Chen et al., 2016)
Peels from pomelo fruit	Single chamber	5.19 mA/cm^2	Threefold compared to plain graphite felt	(Chen et al., 2012c)
Coconut shell + Sewage sludge	Single chamber	969 mW/m^2	2.4-fold higher power density comparing to plain graphite anode	(Yuan et al., 2015)
Bamboo charcoal	Double chamber	1,600 mW/m^2	50% higher comparing to plain graphite tube	(Zhang et al., 2014)
Loofah sponge + TiO$_2$ + carbon	Single chamber	2,590 mW/m^2	201% higher comparing to plain graphite	(Tang et al., 2015)
Loofah sponge + TiO$_2$	Single chamber	1,850 mW/m^2	16% higher comparing to control i.e. Loofa sponge	
Wild mushroom	Single chamber	3.26 mA/cm^2	10.8 higher current density than control	(Karthikeyan et al., 2015)
Corn stem		3.43 mA/cm^2	Eightfold higher current compared to control	
King mushroom		2.93 mA/cm^2	11.4-fold higher power density than control	

differences in physical characters such as porosity and mechanical stability. The summarized power generation noted by using the different natural available materials are presented in Table 5.2.

A few of the easily accessible natural wastes for fabrication of anode material in MFCs includes recycled paper, wood chips, and their resulting wastes; paddy and corn straws; and shells of almond and chestnut. However, in a few countries like Malaysia, it generates excess particular waste like palm oil biomass. These biomasses can be subjected to the fabrication of anode electrodes. In this regard, Huang et al. have analyzed the anode performance by using the waste derived from the coffee beans after the extraction of coffee. In this study, they have noted a maximum current generation of 3,927 mW/m^2. Because of high global coffee consumption, approximately 6 million tons of coffee-based waste is being generated. The use of such wastes, for the fabrication of anodes in MFCs, can decrease the overall cost and can also ease the environmental concerns related to their dumping into food waste (Hung et al., 2019).

Also, many of the waste materials have not been analyzed for the fabrication of anode. Moreover, a few of the earlier electrode materials developed from natural waste are limited by low conductivity, which is one of the crucial parameters that governs MFC performance. As the graphene and their derivates have already proven their conductive capabilities. Therefore, the fabrication of graphene or its derivates using natural waste can be a beneficial approach. Moreover, these derived graphenes from natural waste can further be tailored for carbon-based metal composites to enhance the overall bioelectricity in MFC.

5.3.3 ANODE FABRICATION USING METALS AND METAL OXIDES

Apart from the carbon materials, the frequently used anode materials in MFC include metals. However, the corrosion of metals can render the overall performance of MFCs; also, a few of the metals are non-biocompatible, thereby limiting their scope of application in MFCs. But a few metals like Au, Ni, Ag, Al, Cu, and stainless have demonstrated their applicability in MFCs as functioning anode material. For instance, the use of molybdenum-based material for anodes has exhibited a superior power generation of 1,296 mW/m^2. However, it was limited due to corrosion, during longevity studies (Yamashita & Yokoyama, 2018). Also, the use of porous 3D-based metal oxides like Ti nanosheets have exhibited a better electrochemical kinetic and molecular movement. However, the direct use of metals without modification have been limited with minimal biofilm formation (Yin et al., 2015). Here it should be noted that the metals are often pointed to as highly conductive materials, due to the free movement of outer electrons in the metallic orbitals. But these are often limited with low bioelectrogenic biofilm formation when compared to carbon-based materials.

In order to overcome the shortcomings with metal-based anode materials, alternations of metal surfaces are required. These modifications should retain their conductive nature, along with minimal corrosivity during long-term operation. And also, it should be biocompatible with higher bioelectrogenic biofilm formation. Therefore, several studies have implemented the modification of metals by combining them with other conductive materials like graphene or other carbon derivatives. Also, the surface of metals was altered by oxidation of chemicals, or by treating with ammonia to develop surface charges, that can assist in biofilm formation. Yamashita and Yokoyama have conducted a comprehensive study on the metal-centered anodes in MFCs and have pointed at the use of conductive metals like the Ag, Au, Pt, and Ti can enhance power generation during short-term operation (Yamashita & Yokoyama, 2018). However, these metals are often limited with corrosion in the long run (Little et al., 2020). Moreover, these metals are not economically feasible for the commercial application of MFCs. Therefore, the use of these metals in their oxide forms along with carbon-derivate materials can minimize the fabrication cost of MFC anodes.

5.3.4 IMPROVEMENT IN MFC PERFORMANCE BY ALTERING ANODE SURFACES

The use of composite materials is one of the realistic ways to maximize the MFC output by minimizing existing issues like conductivity, economic feasibility, the readiness of materials, etc. An abundant number of studies have been published on the metal and metal composite-built anodes like a combination of graphite with metal, carbon cloth with metals, and metal alloys and also the use of conductive polymers on the support matrix-like carbon cloth. Various methods like sol-gel process, hydrothermal approach, and solvothermal mixing are the most common processes employed in the preparation of these processes. The performance of these composites as anode material is greatly varied by the synthesis process. For instance, Gong et al. synthesized a $CuCo_2S_4/rGO$ by employing a hydrothermal approach and solvothermal mixing. In this study, they noted a superior performance of $CuCo_2S_4/rGO$ synthesized from the solvothermal procedure (Gong et al., 2018). Similarly, a $Mn^{++}/graphite$ electrode was fabricated by blending manganese sulfate and graphite powder. This composite has exhibited a 1,000-fold increase in performance in comparison to the use of plain graphite (Park & Zeikus, 2002). In further, Fraiwan et al. have noted that the chemical treatment of CNTs has increased their surface activity and also making it easily accessible for bioelectrogenic biofilm formation (Fraiwan et al., 2014). They have also suggested that the pre-treatment of CNTs was a requisite to minimize the activation losses and bacterial toxicity.

The alteration of anode electrodes by the use of conducting polymers has exhibited a better performance in MFCs, which might be due to an increase in bioelectrogenic biofilm and surface modification. The elevated electrical conductivity with outstanding redox characteristics and environmentally friendly nature had made polypyrole (ppy) and poly aniline (PANI) exceptional polymers among all the conducting polymers examined (Ghosh et al., 2020; Li et al., 2011; Mohammadifar et al., 2018). Several studies have accessed these polymers in enhancing the performance of MFCs. However, the use of these polymers on conventional anode materials can increase cost during commercialization. In this regard, Yuan et al. have fabricated a PANI-coated natural loofah sponge anode electrode and noted a maximum power generation of 1.09 W/m^2. This study has provided an effective approach to developing a MFC anode from ecological resources in an economical manner (Yuan et al., 2013). The efficacy of altered electrodes should be tested for long-term operation, and it should be low cost as benchmark. A short summary of different altered electrodes for the anode in MFCs is provided in Table 5.2.

Surface treatment of anodes and electrodes to increase the power generation was initially studied by Cheng and logan. In this study, the carbon cloth electrode was modified by using the ammonia gas along with helium at higher temperatures (700° C). The modified carbon cloth with ammonia gas and control carbon cloth with any pre-treatment has exhibited the highest power density (PD) of 96 W/m³ and 115 W/m³, respectively. Also, the modified carbon cloth exhibited a decrease in the lag period for bioelectrogenic biofilm formation (Cheng & Logan, 2007). The decrease in the lag period of the pre-treated anode is due to the positive surface charge development by treating with ammonia. As most of the bacterial cell membranes are negatively charged, therefore the positively charged anode surface enables a higher biofilm development. By thus, an increase in electron transfer rate kinetics and higher organic reduction can be noted. Moreover, the treatment of anode surface using positively charged gases like ammonia can be implemented to several MFC designs. Also, it should be noted that ammonia treatment enables an increase in power generation during short-term operation. This has yet to be studied in MFC longevity studies for field applications. Moreover, the requirements of high temperatures and optimized complex environment for the surface modification can limit in-field applications.

The modification of anode surface by employing thermal pre-treatment also resulted in a modest increase in power generation in MFC, due to the difficulties in the tuning of material architecture at fixed temperatures. For instance, Wang et al. have investigated the thermally pre-treated carbon mesh as an anode in MFC. In this study, they have noted only a 3% increase in power generation with pre-treated carbon mesh in comparison to the untreated carbon mesh. The thermal pre-treatment of carbon mesh has altered the surface area of the electrode, by achieving a higher bioelectrogenic development. However, it did point to a significant power production (Wang et al., 2009). In comparison to thermal pre-treatment, the acid pre-treatment of the anode is known to be the shortest way to alter the surface morphology of the electrode. In general, the acid pre-treatment is pursued by immersing the electrode in a strongly acidic solution and thereby achieving a protonation on the surface of the electrode. Feng et al. has studied the influence of nitric acid pre-treatment on an anode surface, and have noted a twofold increase in power generation in comparison to control (Feng et al., 2010). In similar to thermal and acid pre-treatment methods, electrochemical oxidation of anodes is another surface alteration strategy. Typically, electrochemical oxidation enables the development of new carboxyl groups, which can assist in bacterial adhesion. The developed carboxyl groups can attach to bioelectrogens by bridging a peptide bond that can aid in electron transfer from cell membrane to the electrode surface. In this regard, Tang et al. have analyzed the power generation using a plain and modified graphite felt. The electrochemical modification of graphite felt was pursued by applying 30 mA/cm² for 12 hours. In this study, they have noted a 40% rise in power generation with modified graphite felt in comparison to control (Tang et al., 2011).

Aside from surface treatment and use of composite material, surface coating of the anode has also exhibited an increase in power generation. The surface coating of the anode has often been described for carbon paper, carbon cloth, metals, composite materials, and CNTs to enhance power generation (Chen et al., 2018a; Liu et al., 2015; Rajesh et al., 2020). The CNTs and other metals attached through coating are the widely used components to modify the electrode surface to increase power generation. These coatings also pointed to a decrease in internal resistance, which is often noted by variation in impedance and cyclic voltammetry analysis. It also noted that the nanoparticles can inhibit bacterial growth. Therefore, further studies are required to analyze the influence of nanoparticles on the growth of bioelectrogens. In further, the anode electrodes are also coated using conductive polymers like Ppy and PANI, which are enabled to achieve a higher electron transfer along with an increase in surface area (Sonawane et al., 2017).

5.4   ECONOMICALLY VIABLE ANODES FABRICATED FROM NATURAL WASTE

As pointed to earlier, anodes are known to be one of the key components in MFCs, as they are in direct contact with bioelectrogens and thereby facilitate energy generation. Nevertheless, the field application of MFCs can be limited by the economic issues related to anode modification and fabrication. Therefore, the fabrication of anodes using waste can be beneficial. Moreover, a few researchers have pointed out that the cost of electrodes in over cost of MFCs can vary from 20–50% on the basis of material and fabrication type. On the basis of cost evaluations made by Li et al., the conventionally used anode materials like GAC and graphite flakes cost around \$1,250 to \$1,450 per ton (Li et al., 2017b). This is significantly higher in comparison to carbon paper and carbon cloth, which generally costs around \$100,000 to \$500,000 m^2 (Huggins et al., 2014). Nevertheless, the use of GAC and graphite flakes still is an unviable option due to washout and the filled anode chambers using GAC can be blocked during the continuous mode operation.

In this context, the development of anode materials like biochar, by using economically viable, readily available nonrenewable sources such as biomass, points to a viable option. Therefore, various research groups have demonstrated the feasibility of employing biomass as a precursor material to build anode fabrication. In this regard, the precursor materials like sewage sludge mixed fly ash, corn stem, cardboard, palm kernel shell, etc., are tested for biochar generation and further to be used as the anode. Also, the conductivity of these biochar materials can further be tuned by mixing with conductive polymers and metal oxides. Table. 5.3 summarizes the power generations and the cost of economically viable anodes fabricated from natural waste. The performance of these materials is found to be as similar to conventional GAC and graphite flakes. As there is only a limited literature available in comparison of conventional materials with biochar, therefore further studies are needed to evaluate the biochar as an economical viable or an unfeasible option.

TABLE 5.3
Summarized Consumer Prices for a Few of the Conventional Anode Materials Used in MFCs

Electrode material	Price per ton (US $)
Cu	5,528
Stainless steel	2,645
Nickel	14,898
Graphite-natural large flakes	1,450
Activated carbon	1,250
Platinum	3.6×10^7
Graphene	2.0×10^5
Graphene, nano-powder	2.4×10^6
Single-layer graphene powder	1.0×10^6
Multi-walled carbon nanotube	1.9×10^6
500 nm length multiwalled carbon nanotube	7.0×10^5
Biochar from compressed milling waste	381
Biochar from forestry waste	381
Granular-activated carbon	2,500
Graphite granules	800

Adapted from Li et al. (2017b) and Yaqoob et al. (2020a).

5.5 CONCLUSION

One of the greatest issues in MFCs is the proper assortment of materials for anode fabrication. Based on earlier studies, the anode materials derived from the wastes and from other nonrenewable sources seem to be practical and more well suited than the traditional materials that are often used in MFC. As biomass-derived materials are proved to be cheaper, however, they are limited in energy generation due to poor conductivity. Therefore, it can limit their practical applicability in bioelectricity generation with concurrent treatment of organic reduction in waste. In order to tackle this problem, the anode materials developed from biomass should be reworked by incorporating the metals and metal oxides or coating of these materials with conductive polymers. Consequently, significant investigations regarding the biomass-derived anodes are essential to optimize the functioning of MFCs on a profitable scale.

REFERENCES

Adachi, M., Shimomura, T., Komatsu, M., Yakuwa, H., & Miya, A. (2008). A novel mediator–polymer-modified anode for microbial fuel cells. *Chemical Communications*, 17, 2055–2057.

Borsje, C., Liu, D., Sleutels, T. H. J. A., Buisman, C. J. N., & Ter Heijne, A. (2016). Performance of single carbon granules as perspective for larger scale capacitive bioanodes. *Journal of Power Sources, 325*, 690–696.

Cai, T., Meng, L., Chen, G., Xi, Y., Jiang, N., Song, J., Zheng, S., Liu, Y., Zhen, G., & Huang, M. (2020). Application of advanced anodes in microbial fuel cells for power generation: A review. *Chemosphere, 248*, 125985.

Caizán-Juanarena, L., Servin-Balderas, I., Chen, X., Buisman, C. J. N., & Ter Heijne, A. (2019). Electrochemical and microbiological characterization of single carbon granules in a multi-anode microbial fuel cell. *Journal of Power Sources, 435*, 126514.

Chandrasekhar, K. (2019). Effective and nonprecious cathode catalysts for oxygen reduction reaction in microbial fuel cells. In: S. V. Mohan, S. Varjani, & A. Pandey (Eds.), *Microbial electrochemical technology* (pp. 485–501). Elsevier.

Chandrasekhar, K., & Ahn, Y. H. (2017). Effectiveness of piggery waste treatment using microbial fuel cells coupled with elutriated-phased acid fermentation. *Bioresource Technology, 244*(Pt 1), 650–657.

Chandrasekhar, K., Cayetano, R. D. A., Ikram, M., Kumar, G., & Kim, S.-H. (2020a). Evaluation of the biochemical methane potential of different sorts of Algerian date biomass. *Environmental Technology & Innovation, 20*, 101180.

Chandrasekhar, K., Kumar, A. N., Raj, T., Kumar, G., & Kim, S.-H. (2021a). Bioelectrochemical system-mediated waste valorization. *Systems Microbiology and Biomanufacturing, 1*, 432–443

Chandrasekhar, K., Kumar, G., Venkata Mohan, S., Pandey, A., Jeon, B.-H., Jang, M., & Kim, S. H. (2020b). Microbial electro-remediation (MER) of hazardous waste in aid of sustainable energy generation and resource recovery. *Environmental Technology & Innovation, 19*, 100997.

Chandrasekhar, K., Mehrez, I., Kumar, G., & Kim, S.-H. (2021b). Relative evaluation of acid, alkali, and hydrothermal pre-treatment influence on biochemical methane potential of date biomass. *Journal of Environmental Chemical Engineering, 9*(5), 106031.

Chandrasekhar, K., Naresh Kumar, A., Kumar, G., Kim, D.-H., Song, Y.-C., & Kim, S.-H. (2021c). Electro-fermentation for biofuels and biochemicals production: Current status and future directions. *Bioresource Technology, 323*, 124598.

Chandrasekhar, K., Velvizhi, G., & Venkata Mohan, S. (2021d). Bio-electrocatalytic remediation of hydrocarbons contaminated soil with integrated natural attenuation and chemical oxidant. *Chemosphere, 280*, 130649.

Chandrasekhar, K., & Venkata Mohan, S. (2012). Bio-electrochemical remediation of real field petroleum sludge as an electron donor with simultaneous power generation facilitates biotransformation of PAH: Effect of substrate concentration. *Bioresource Technology, 110*, 517–525.

Chandrasekhar, K., & Venkata Mohan, S. (2014a). Bio-electrohydrolysis as a pre-treatment strategy to catabolize complex food waste in closed circuitry: Function of electron flux to enhance acidogenic biohydrogen production. *International Journal of Hydrogen Energy, 39*(22), 11411–11422.

Chandrasekhar, K., & Venkata Mohan, S. (2014b). Induced catabolic bio-electrohydrolysis of complex food waste by regulating external resistance for enhancing acidogenic biohydrogen production. *Bioresource Technology, 165*, 372–382.

Chen, B.-Y., Tsao, Y.-T., & Chang, S.-H. (2018a). Cost-effective surface modification of carbon cloth electrodes for microbial fuel cells by candle soot coating. *Coatings, 8*(12), 468.

Chen, Q., Pu, W., Hou, H., Hu, J., Liu, B., Li, J., Cheng, K., Huang, L., Yuan, X., Yang, C., & Yang, J. (2018b). Activated microporous-mesoporous carbon derived from chestnut shell as a sustainable anode material for high performance microbial fuel cells. *Bioresource Technology*, *249*, 567–573.

Chen, S., He, G., Hu, X., Xie, M., Wang, S., Zeng, D., Hou, H., & Schröder, U. (2012a). A three-dimensionally ordered macroporous carbon derived from a natural resource as anode for microbial bioelectrochemical systems. *ChemSusChem*, *5*(6), 1059–1063.

Chen, S., He, G., Liu, Q., Harnisch, F., Zhou, Y., Chen, Y., Hanif, M., Wang, S., Peng, X., Hou, H., & Schröder, U. (2012b). Layered corrugated electrode macrostructures boost microbial bioelectrocatalysis. *Energy & Environmental Science*, *5*(12), 9769–9772.

Chen, S., Liu, Q., He, G., Zhou, Y., Hanif, M., Peng, X., Wang, S., & Hou, H. (2012c). Reticulated carbon foam derived from a sponge-like natural product as a high-performance anode in microbial fuel cells. *Journal of Materials Chemistry*, *22*(35), 18609–18613.

Chen, S., Tang, J., Jing, X., Liu, Y., Yuan, Y., & Zhou, S. (2016). A hierarchically structured urchin-like anode derived from chestnut shells for microbial energy harvesting. *Electrochimica Acta*, *212*, 883–889.

Chen, W., Huang, Y.-X., Li, D.-B., Yu, H.-Q., & Yan, L. (2014). Preparation of a macroporous flexible three dimensional graphene sponge using an ice-template as the anode material for microbial fuel cells. *RSC Advances*, *4*(41), 21619–21624.

Cheng, S., & Logan, B. E. (2007). Ammonia treatment of carbon cloth anodes to enhance power generation of microbial fuel cells. *Electrochemistry Communications*, *9*(3), 492–496.

Deval, A. S., Parikh, H. A., Kadier, A., Chandrasekhar, K., Bhagwat, A. M., & Dikshit, A. K. (2017). Sequential microbial activities mediated bioelectricity production from distillery wastewater using bio-electrochemical system with simultaneous waste remediation. *International Journal of Hydrogen Energy*, *42*(2), 1130–1141.

Enamala, M. K., Dixit, R., Tangellapally, A., Singh, M., Dinakarrao, S. M. P., Chavali, M., Pamanji, S. R., Ashokkumar, V., Kadier, A., & Chandrasekhar, K. (2020). Photosynthetic microorganisms (Algae) mediated bioelectricity generation in microbial fuel cell: Concise review. *Environmental Technology & Innovation*, *19*, 100959.

Feng, Y., Yang, Q., Wang, X., & Logan, B. E. (2010). Treatment of carbon fiber brush anodes for improving power generation in air–cathode microbial fuel cells. *Journal of Power Sources*, *195*(7), 1841–1844.

Fraiwan, A., Adusumilli, S. P., Han, D., Steckl, A. J., Call, D. F., Westgate, C. R., & Choi, S. (2014). Microbial power-generating capabilities on micro-/nano-structured anodes in micro-sized microbial fuel cells *Fuel Cells*, *14*(6), 801–809.

Gambino, E., Chandrasekhar, K., & Nastro, R. A. (2021). SMFC as a tool for the removal of hydrocarbons and metals in the marine environment: A concise research update. *Environmental Science and Pollution Research*, *28*(24), 30436–30451.

Ghosh, S., Das, S., & Mosquera, M. E. G. (2020). Conducting polymer-based nanohybrids for fuel cell application. *Polymers*, *12*(12), 2993.

Gong, Y., Zhao, J., Wang, H., & Xu, J. (2018). CuCo$_2$S$_4$/reduced graphene oxide nanocomposites synthesized by one-step solvothermal method as anode materials for sodium ion batteries. *Electrochimica Acta*, *292*, 895–902.

Guo, W., Pi, Y., Song, H., Tang, W., & Sun, J. (2012). Layer-by-layer assembled gold nanoparticles modified anode and its application in microbial fuel cells. *Colloids and Surfaces A: Physicochemical and Engineering Aspects*, *415*, 105–111.

Hassan, S. H. A., Kim, Y. S., & Oh, S.-E. (2012). Power generation from cellulose using mixed and pure cultures of cellulose-degrading bacteria in a microbial fuel cell. *Enzyme and Microbial Technology*, *51*(5), 269–273.

He, Y., Liu, Z., Xing, X.-H., Li, B., Zhang, Y., Shen, R., Zhu, Z., & Duan, N. (2015). Carbon nanotubes simultaneously as the anode and microbial carrier for up-flow fixed-bed microbial fuel cell. *Biochemical Engineering Journal*, *94*, 39–44.

Hindatu, Y., Annuar, M. S. M., & Gumel, A. M. (2017). Mini-review: Anode modification for improved performance of microbial fuel cell. *Renewable and Sustainable Energy Reviews*, *73*, 236–248.

Hu, M., Li, X., Xiong, J., Zeng, L., Huang, Y., Wu, Y., Cao, G., & Li, W. (2019). Nano-$Fe_3C@PGC$ as a novel low-cost anode electrocatalyst for superior performance microbial fuel cells. *Biosensors and Bioelectronics*, *142*, 111594.

Huggins, T., Wang, H., Kearns, J., Jenkins, P., & Ren, Z. J. (2014). Biochar as a sustainable electrode material for electricity production in microbial fuel cells. *Bioresource Technology*, *157*, 114–119.

Hung, Y.-H., Liu, T.-Y., & Chen, H.-Y. (2019). Renewable coffee waste-derived porous carbons as anode materials for high-performance sustainable microbial fuel cells. *ACS Sustainable Chemistry & Engineering*, *7*(20), 16991–16999.

Kamedulski, P., Ilnicka, A., Lukaszewicz, J. P., & Skorupska, M. (2019). Highly effective three-dimensional functionalization of graphite to graphene by wet chemical exfoliation methods. *Adsorption*, *25*(3), 631–638.

Karthikeyan, R., Wang, B., Xuan, J., Wong, J. W. C., Lee, P. K. H., & Leung, M. K. H. (2015). Interfacial electron transfer and bioelectrocatalysis of carbonized plant material as effective anode of microbial fuel cell. *Electrochimica Acta*, *157*, 314–323.

Kondaveeti, S., Choi, K. S., Kakarla, R., & Min, B. (2014a). Microalgae Scenedesmus obliquus as renewable biomass feedstock for electricity generation in microbial fuel cells (MFCs). *Frontiers of Environmental Science & Engineering*, *8*(5), 784–791.

Kondaveeti, S., Lee, J., Kakarla, R., Kim, H. S., & Min, B. (2014b). Low-cost separators for enhanced power production and field application of microbial fuel cells (MFCs). *Electrochimica Acta*, *132*(0), 434–440.

Kondaveeti, S. K., Seelam, J. S., & Mohanakrishna, G. (2018). Anodic electron transfer mechanism in bioelectrochemical systems. In: D. Das (Ed.), *Microbial fuel cell: A bioelectrochemical system that converts waste to Watts* (pp. 87–100). Springer International Publishing, Cham.

Kong, X., Zhu, Y., Lei, H., Wang, C., Zhao, Y., Huo, E., Lin, X., Zhang, Q., Qian, M., Mateo, W., Zou, R., Fang, Z., & Ruan, R. (2020). Synthesis of graphene-like carbon from biomass pyrolysis and its applications. *Chemical Engineering Journal*, *399*, 125808.

Kumar, A. K., Reddy, M. V., Chandrasekhar, K., Srikanth, S., & Mohan, S. V. (2012). Endocrine disruptive estrogens role in electron transfer: bio-electrochemical remediation with microbial mediated electrogenesis. *Bioresource Technology*, *104*, 547–556.

Li, C., Zhang, L., Ding, L., Ren, H., & Cui, H. (2011). Effect of conductive polymers coated anode on the performance of microbial fuel cells (MFCs) and its biodiversity analysis. *Biosensors and Bioelectronics*, *26*(10), 4169–4176.

Li, D., Deng, L., Yuan, H., Dong, G., Chen, J., Zhang, X., Chen, Y., & Yuan, Y. (2018). N, P-doped mesoporous carbon from onion as trifunctional metal-free electrode modifier for enhanced power performance and capacitive manner of microbial fuel cells. *Electrochimica Acta*, *262*, 297–305.

Li, F., Yarnykh, V. L., Hatsukami, T. S., Chu, B., Balu, N., Wang, J., Underhill, H. R., Zhao, X., Smith, R., & Yuan, C. (2010). Scan-rescan reproducibility of carotid atherosclerotic plaque morphology and tissue composition measurements using multicontrast MRI at 3T. *Journal of Magnetic Resonance Imaging, 31*(1), 168–176.

Li, J., Li, H., Fu, Q., Liao, Q., Zhu, X., Kobayashi, H., & Ye, D. (2017a). Voltage reversal causes bioanode corrosion in microbial fuel cell stacks. *International Journal of Hydrogen Energy, 42*(45), 27649–27656.

Li, S., Cheng, C., & Thomas, A. (2017b). Carbon-based microbial-fuel-cell electrodes: From conductive supports to active catalysts. *Advanced Materials, 29*(8), 1602547.

Little, B. J., Blackwood, D. J., Hinks, J., Lauro, F. M., Marsili, E., Okamoto, A., Rice, S. A., Wade, S. A., & Flemming, H. C. (2020). Microbially influenced corrosion—Any progress? *Corrosion Science, 170*, 108641.

Liu, H., Ramnarayanan, R., & Logan, B. E. (2004). Production of electricity during wastewater treatment using a single chamber microbial fuel cell. *Environmental Science & Technology, 38*(7), 2281–2285.

Liu, J., Liu, J., He, W., Qu, Y., Ren, N., & Feng, Y. (2014). Enhanced electricity generation for microbial fuel cell by using electrochemical oxidation to modify carbon cloth anode. *Journal of Power Sources, 265*, 391–396.

Liu, J., Qiao, Y., Guo, C. X., Lim, S., Song, H., & Li, C. M. (2012). Graphene/carbon cloth anode for high-performance mediatorless microbial fuel cells. *Bioresource Technology, 114*, 275–280.

Liu, M., Zhou, M., Yang, H., Zhao, Y., & Hu, Y. (2015). A cost-effective polyurethane based activated carbon sponge anode for high-performance microbial fuel cells. *RSC Advances, 5*(102), 84269–84275.

Logan, B., Cheng, S., Watson, V., & Estadt, G. (2007). Graphite fiber brush anodes for increased power production in air-cathode microbial fuel cells. *Environmental Science & Technology, 41*(9), 3341–3346.

Mal, N., Srivastava, K., Sharma, Y., Singh, M., Rao, K. M., Enamala, M. K., Chandrasekhar, K., & Chavali, M. (2021). Facets of diatom biology and their potential applications. *Biomass Conversion and Biorefinery*, doi: https://doi.org/10.1007/s13399-020-01155-5

Mehdinia, A., Ziaei, E., & Jabbari, A. (2014). Facile microwave-assisted synthesized reduced graphene oxide/tin oxide nanocomposite and using as anode material of microbial fuel cell to improve power generation. *International Journal of Hydrogen Energy, 39*(20), 10724–10730.

Mehranian, A., Ay, M. R., Alam, N. R., & Zaidi, H. (2010). Quantifying the effect of anode surface roughness on diagnostic x-ray spectra using Monte Carlo simulation. *Medical Physics, 37*(2), 742–752.

Mohammadifar, M., Yazgan, I., Zhang, J., Kariuki, V., Sadik, O. A., & Choi, S. (2018). Green biobatteries: Hybrid paper–polymer microbial fuel cells. *Advanced Sustainable Systems, 2*(10), 1800041.

Mohan, S. V., & Chandrasekhar, K. (2011a). Self-induced bio-potential and graphite electron accepting conditions enhances petroleum sludge degradation in bio-electrochemical system with simultaneous power generation. *Bioresource Technology, 102*(20), 9532–9541.

Mohan, S. V., & Chandrasekhar, K. (2011b). Solid phase microbial fuel cell (SMFC) for harnessing bioelectricity from composite food waste fermentation: influence of electrode assembly and buffering capacity. *Bioresource Technology, 102*(14), 7077–7085.

Mohanakrishna, G., Abu-Reesh, I. M., Kondaveeti, S., Al-Raoush, R. I., & He, Z. (2018). Enhanced treatment of petroleum refinery wastewater by short-term applied voltage in single chamber microbial fuel cell. *Bioresource Technology, 253*, 16–21.

Park, D., & Zeikus, J. (2002). Impact of electrode composition on electricity generation in a single-compartment fuel cell using *Shewanella putrefaciens*. *Applied Microbiology and Biotechnology, 59*(1), 58–61.

Park, J.-H., Chandrasekhar, K., Jeon, B.-H., Jang, M., Liu, Y., & Kim, S.-H. (2021). State-of-the-art technologies for continuous high-rate biohydrogen production. *Bioresource Technology, 320*, 124304.

Purkait, T., Singh, G., Singh, M., Kumar, D., & Dey, R. S. (2017). Large area few-layer graphene with scalable preparation from waste biomass for high-performance supercapacitor. *Scientific Reports, 7*(1), 15239.

Quan, X., Sun, B., & Xu, H. (2015). Anode decoration with biogenic Pd nanoparticles improved power generation in microbial fuel cells. *Electrochimica Acta, 182*, 815–820.

Raj, T., Chandrasekhar, K., Banu, R., Yoon, J.-J., Kumar, G., & Kim, S.-H. (2021a). Synthesis of γ-valerolactone (GVL) and their applications for lignocellulosic deconstruction for sustainable green biorefineries. *Fuel, 303*, 121333.

Raj, T., Chandrasekhar, K., Kumar, A. N., & Kim, S.-H. (2021b). Recent biotechnological trends in lactic acid bacterial fermentation for food processing industries. *Systems Microbiology and Biomanufacturing, 1*, 14–40.

Rajesh, P. P., Noori, M. T., & Ghangrekar, M. M. (2020). Improving performance of microbial fuel cell by using polyaniline-coated carbon–felt anode. *Journal of Hazardous, Toxic, and Radioactive Waste, 24*(3), 04020024.

Sarathi, V. S., & Nahm, K. S. (2013). Recent advances and challenges in the anode architecture and their modifications for the applications of microbial fuel cells. *Biosensors and Bioelectronics, 43*, 461–475.

Sonawane, J. M., Yadav, A., Ghosh, P. C., & Adeloju, S. B. (2017). Recent advances in the development and utilization of modern anode materials for high performance microbial fuel cells. *Biosensors and Bioelectronics, 90*, 558–576.

Sun, Y., Wei, J., Liang, P., & Huang, X. (2011). Electricity generation and microbial community changes in microbial fuel cells packed with different anodic materials. *Bioresource Technology, 102*(23), 10886–10891.

Tang, J., Yuan, Y., Liu, T., & Zhou, S. (2015). High-capacity carbon-coated titanium dioxide core–shell nanoparticles modified three dimensional anodes for improved energy output in microbial fuel cells. *Journal of Power Sources, 274*, 170–176.

Tang, X., Guo, K., Li, H., Du, Z., & Tian, J. (2011). Electrochemical treatment of graphite to enhance electron transfer from bacteria to electrodes. *Bioresource Technology, 102*(3), 3558–3560.

ter Heijne, A., Hamelers, H. V. M., Saakes, M., & Buisman, C. J. N. (2008). Performance of non-porous graphite and titanium-based anodes in microbial fuel cells. *Electrochimica Acta, 53*(18), 5697–5703.

Wang, K., Liu, Y., & Chen, S. (2011). Improved microbial electrocatalysis with neutral red immobilized electrode. *Journal of Power Sources, 196*(1), 164–168.

Wang, X., Cheng, S., Feng, Y., Merrill, M. D., Saito, T., & Logan, B. E. (2009). Use of carbon mesh anodes and the effect of different pre-treatment methods on power production in microbial fuel cells. *Environmental Science & Technology, 43*(17), 6870–6874.

Wei, J., Liang, P.,& Huang, X. (2011). Recent progress in electrodes for microbial fuel cells. *Bioresource Technology, 102*(20), 9335–9344.

Xie, X., Hu, L., Pasta, M., Wells, G. F., Kong, D., Criddle, C. S., & Cui, Y. (2011). Three-dimensional carbon nanotube–textile anode for high-performance microbial fuel cells. *Nano Letters, 11*(1), 291–296.

Yamashita, T., & Yokoyama, H. (2018). Molybdenum anode: A novel electrode for enhanced power generation in microbial fuel cells, identified via extensive screening of metal electrodes. *Biotechnology for Biofuels*, *11*(1), 39.

Yaqoob, A. A., Ibrahim, M. N. M., & Rodríguez-Couto, S. (2020a). Development and modification of materials to build cost-effective anodes for microbial fuel cells (MFCs): An overview. *Biochemical Engineering Journal*, *164*, 107779.

Yaqoob, A. A., Umar, K., & Ibrahim, M. N. M. (2020b). Silver nanoparticles: Various methods of synthesis, size affecting factors and their potential applications–a review. *Applied Nanoscience*, *10*(5), 1369–1378.

Yin, T., Lin, Z., Su, L., Yuan, C., & Fu, D. (2015). Preparation of vertically oriented TiO2 nanosheets modified carbon paper electrode and its enhancement to the performance of MFCs. *ACS Applied Materials & Interfaces*, *7*(1), 400–408.

Yuan, Y., Jeon, Y., Ahmed, J., Park, W., & Kim, S. (2009). Use of carbon nanoparticles for bacteria immobilization in microbial fuel cells for high power output. *Journal of The Electrochemical Society*, *156*(10), B1238.

Yuan, Y., & Kim, S.-H. (2008). Polypyrrole-coated reticulated vitreous carbon as anode in microbial fuel cell for higher energy output. *Bulletin of the Korean Chemical Society*, *29*(1), 168–172.

Yuan, Y., Liu, T., Fu, P., Tang, J., & Zhou, S. (2015). Conversion of sewage sludge into high-performance bifunctional electrode materials for microbial energy harvesting. *Journal of Materials Chemistry A*, *3*(16), 8475–8482.

Yuan, Y., Zhou, S., Liu, Y., & Tang, J. (2013). Nanostructured macroporous bioanode based on polyaniline-modified natural loofah sponge for high-performance microbial fuel cells. *Environmental Science & Technology*, *47*(24), 14525–14532.

Yuan, Y., Zhou, S., Zhao, B., Zhuang, L., & Wang, Y. (2012). Microbially-reduced graphene scaffolds to facilitate extracellular electron transfer in microbial fuel cells. *Bioresource Technology*, *116*, 453–458.

Zhang, J., Li, J., Ye, D., Zhu, X., Liao, Q., & Zhang, B. (2014). Tubular bamboo charcoal for anode in microbial fuel cells. *Journal of Power Sources*, *272*, 277–282.

Zhang, S., Wang, H., Liu, J., & Bao, C. (2020). Measuring the specific surface area of monolayer graphene oxide in water. *Materials Letters*, *261*, 127098.

Zhang, T., Zeng, Y., Chen, S., Ai, X., & Yang, H. (2007). Improved performances of E. coli-catalyzed microbial fuel cells with composite graphite/PTFE anodes. *Electrochemistry Communications*, *9*(3), 349–353.

Zhang, Y., Liu, L., Van der Bruggen, B., & Yang, F. (2017). Nanocarbon based composite electrodes and their application in microbial fuel cells. *Journal of Materials Chemistry A*, *5*(25), 12673–12698.

Zhang, Y., Sun, J., Hu, Y., Li, S., & Xu, Q. (2013). Carbon nanotube-coated stainless steel mesh for enhanced oxygen reduction in biocathode microbial fuel cells. *Journal of Power Sources*, *239*, 169–174.

Zhao, Y., Watanabe, K., Nakamura, R., Mori, S., Liu, H., Ishii, K., & Hashimoto, K. (2010). Three-dimensional conductive nanowire networks for maximizing anode performance in microbial fuel cells. *Chemistry – A European Journal*, *16*(17), 4982–4985.

Zhong, D., Liao, X., Liu, Y., Zhong, N., & Xu, Y. (2018). Enhanced electricity generation performance and dye wastewater degradation of microbial fuel cell by using a petaline NiO@ polyaniline-carbon felt anode. *Bioresource Technology*, *258*, 125–134.

6 Electro-Fermentation Technology: Synthesis of Chemicals and Biofuels

Devashish Tribhuvan, Vinay V, Saurav Gite, and Shadab Ahmed
Department of Biotechnology (merged with Institute of Bioinformatics and Biotechnology), Savitribai Phule Pune University, Pune, India

CONTENTS

DOI: 10.1201/9781003225430-6

6.1   INTRODUCTION

The rapid rates of industrialization and urbanization across the world, combined with diminishing energy resources, have further created a big gap in the balance between energy supply and demands. And this has also accompanied the growing environmental problem of waste management (Ganbino et al., 2021; Venkata Mohan et al., 2011). Now in this scenario, more and more pollutants in the form of hazardous toxic chemicals and organic wastes are being released into the environment and the treatment plant potential is far from matching up with the waste generated (Venkata Mohan et al., 2013; Kadier et al., 2017). With the above background, researchers all over the world have started paying increasing attention towards advances and efficient technology development for sustainable bioprocesses directed towards the synthesis of various chemical compounds and biofuels by renewable means (Lee et al., 2015; H. S. Lee et al., 2016; J. Y. Lee et al., 2016; Kumar et al., 2018; Bakonyi et al., 2018).

The traditional fermentation techniques typically employ fairly pure substrates and that largely increases the production expense (Enamala et al., 2019; Lee et al., 2011; Sindhu et al., 2011; Cok et al., 2014; Roy et al., 2016). Secondly, very precise use of culture media for a particular microbial fermentation process, which includes vitamins, minerals, nitrogen sources, antifoaming agents, chelating factors, and buffers, can substantially add to the expense of production (Raj et al., 2021a, 2021b). Also, the yield and purity obtained in a traditional fermentation can be a restraining aspect, because of an imbalance in the metabolic pathway (Schievano et al., 2016; Bursac et al., 2017). The micro-bubbling in traditional fermentation causes variation in mass-transfer efficiency and so to improve gas-phase solubilization in the bioreactor, it needs to be pressurized, which again increases reactor engineering costs (Chandrasekhar et al., 2021b; Soccol et al., 2013).

Electro-fermentation involves the microbial transformation of organic molecules, just like the traditional method, but the process is facilitated by mediated electron transfer (MET), where suitable electrodes serve as ports for electron donation-acceptance (Nastro et al., 2014; Roy et al., 2016; Endreny et al., 2020; Choi & Sang, 2016; Nastro, 2014). This helps to regulate redox balance, pH, thermodynamics, and further metabolic rates or flux towards the targeted product. The EF, a hybrid technique, is also economical, lucrative, efficient, and environmentally friendly compared to conventional fermentation (Chandrasekhar et al., 2014; Schievano et al., 2016).

Based on the electroactive microorganism utilized, EF can be subdivided into the following:

a. Microbial fuel cell (MFC),
b. Microbial electrolysis cell (MEC),
c. Bioelectrochemical system (BES), and
d. Microbial desalination system (MDS).

The most basic EF setup is a MFC where organic compounds are acted upon by microbes and the ensuing bioelectrochemical reactions produce electrons that can be used to generate electricity. If this setup is slightly modified to provide external current to MFCs, then it reduces the substrate present at the cathode to form a product; therefore, operating like a microbial electrolysis cell (MEC). Similarly, many other customization possibilities for a specific motive do exist but all share a general underlying principle. These cell compartments can be single or multiple chambered or even of specific shape like tubular, baffled, staked, etc., catering to the requirement. These modifications enhance the utility and efficiency of the system (Kumar et al., 2018).

From the last decade, EF using renewable resources has made use of the synthesis of value-added chemical compounds such as bioalcohols (butanol), diols (1,3-propanediol), carboxylic acids (short-chain and medium-chain fatty acids), amino acids, and many different biofuels (Chandrasekhar et al., 2015; Liu et al., 2019). It has also been reported that food waste can serve as a good substrate for electroactive microbes to oxidize organic and inorganic acids to specific value-added chemicals, thereby contributing towards solid waste management (Chandrasekhar et al., 2015; Liu et al., 2019). If such a strategy is successfully scaled up to industrial levels, it surely can provide a huge incentive for solid waste management.

Herein this chapter, the focus is placed on the mechanisms of electro-fermentation, microbe-electrode interactions, and substrates, and finally the processes involved in the synthesis of different chemicals and biofuels.

6.2 ELECTRO-FERMENTATION

6.2.1 Principle

In the EF process, the limitations of traditional fermentation strategies, namely having low yield and bioconversion efficiency, are surmounted with the stationing of electrodes in the medium that in turn enhances the efficiency of the bioelectrochemical reaction (Chandrasekhar et al., 2021a). Typically, electro-fermentation uses non-depleted electrodes as electron donors or acceptors, which helps to electrochemically regulate the fermentative metabolism of self-derived microbial enzymes (Moscoviz et al., 2016; Schievano et al., 2016). Similarly, it is also verified that making use of such electrodes in EF did provide means to regulate the cellular redox metabolism, which in turn increased the production of target compounds with minimal side products (Kracke et al., 2018).

EF makes use of specialized electrodes (anode and cathode) with improved capability to accept or release electrons during microbial fermentation. In electro-fermentation, self-sustaining anodic microbes oxidize biological substrates and produce electrons and, at the same time, cathodic microbes receive these electrons, which reduces the substrate present in the cathodic chamber, ultimately producing desired chemicals and/or biofuel. This whole process of the exchange of electrons between the electrodes is mediated by specific ion-exchange membranes (Bhagchandanii et al., 2020). Therefore, the efficiency and performance of an EF system are solely determined by the bioelectrochemical activity of the selected microbes for such a fermentation process. By using suitable electrode material and electroactive microbes, one can increase the efficiency of the EF system, resulting in enhancing the productivity of specific chemicals (Choi & Sang, 2016).

6.2.2 Working Mechanism of EF

The most important aspect of the EF technology involves the presence of a double chamber, one of which is the anode chamber and the other is the cathode chamber, respectively, parted by a very specific and selective ion-exchange membrane that can be visualized in Figure 6.1a and b (Bhagchandanii et al., 2020).

6.2.2.1 Bioelectrochemical Reactions at the Anodic and Cathodic Chambers

A typical anodic chamber setup includes components like a suitable electro-chemically active bacteria, an anode electrode, organic matter, and different electrolytes. It is the bioconversion of organic matter by the electrochemical active microbe that produces electrons and protons in this chamber. Further, these electrons get transferred from microbes to the anode by the mechanism of extracellular electron transfer (EET) (Figure 6.1a and b). Further, the EET may involve either a direct or an indirect electron transfer mechanism. In the previous method, microbes directly get in touch with the electrodes and transfer electrons, while the latter method requires the help of third-party electron transfer intermediaries (e.g., neutral red, thionin, riboflavin, etc.) (Choi & Sang, 2016; Velvizhi & Venkata Mohan, 2015). Thereafter, electrons produced at an anode electrode get displaced to the cathode electrode, bypassing an ion/proton exchange membrane (PEM) (Venkata Mohan & Chandrasekhar, 2011a; Venkata Mohan & Chandrasekhar, 2011b; Chandrasekhar & Venkata Mohan, 2012). Ultimately, the electron from the cathode electrode gets transferred to the microbes (Figure 6.1a and b). The

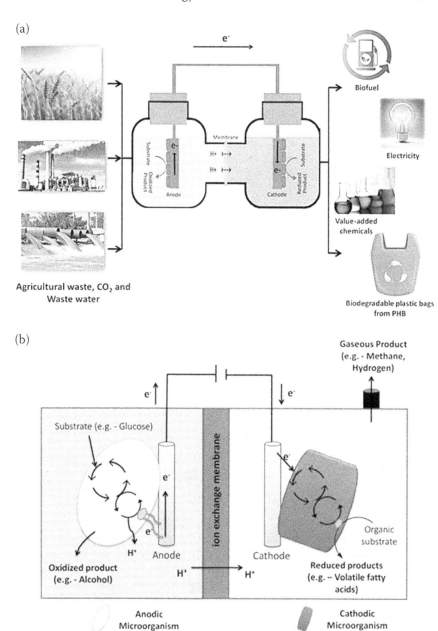

FIGURE 6.1 (a) Schematic representation of electro-fermentation; (b) general scheme of electro-fermentation (EF).

previously mentioned electron transfer process catalyzes reduction processes and produces various value-added chemicals such as ethanol (C_2H_5OH), hydrogen (H_2), methane (CH_4), carboxylic acid, biopolymers, and biofuels at the cathodic compartment (Figure 6.1). These bioelectrochemical reactions in the fermentative pathway are sensitive to changes in pH, temperature, inoculum size, and media composition because the eminence and yield of the product are affected. So it is necessary to maintain optimized conditions during EF to sustain the required oxidation-reduction potential (ORP) during the fermentation process, even if some additives may need to be added to the production medium (Lovley & Holmes, 2021).

6.3   ELECTRODE MATERIAL

6.3.1   ANODE MATERIAL

In electro-fermentation, the anode electrode surface is the place where electrons are accepted. For more yield of product, there should be good interaction between microbes and anodes. So, for the more effective workings of the anode electrode, anode material should have the following necessary distinguishing features, such as high electrical conductivity, non-corrosiveness, biocompatibility, large surface area, high porosity, good chemical stability, high mechanical strength, environmentally friendly, and material should be economical (Wei et al., 2011; Santoro et al., 2017; Bhagchandanii et al., 2020; Chandrasekhar et al., 2021c). Based on the properties, a few materials have been developed to achieve the maximum output.

6.3.1.1   Carbon-Based Materials

According to anodic electrode arrangement, every organic material falls into one of three categories: plane, packed, or brush structure.

6.3.1.1.1   Plane Structure

Plane electrodes make use of graphite sheets, graphite plates, carbon paper, carbon cloth, and carbon mesh, generally as materials for the anode electrode. Carbon paper and carbon cloth are relatively more spongy and have a larger surface area but are expensive, whereas, graphite plates and sheets are cheap but offer good electrical conductivity with lower available surface area for the electron transfer and so produce less product yield (Wei et al. 2011).

6.3.1.1.2   Packed Structure

These are made up of granular and irregular graphite, carbon or graphite felt, or granular-activated carbon (GAC). The plus point of this type of packed structure is that they have a very high surface area for microbial bioelectrochemical interactions. But, these are mainly utilized as packing material, while not generally as a sole individual anode as the available surface area is in the nanoscale range.

6.3.1.1.3   Brush Structure

The brush structures possess high porosity and surface area and are typically made up of carbon brush or graphite brush. In these structures, titanium may be used as a core in which carbon fibers are arranged in a twisted manner.

6.3.1.2   Metal-Based Materials

In general, it has been observed that metals like special-grade titanium and stainless steel have become the first-choice materials because of their high conductivity and non-corrosiveness, although reports also suggest the use of copper, nickel, and silver as electrode material (Tang et al., 2021; Logan et al., 2019). Not only this but even gold (Au) can serve as excellent electrode material in some cases; for example, *Geobacter sulfurreducens* can grow and transfer electrons to gold as effectively as in graphite anode (Richter et al., 2008).

6.3.1.3   Surface Modification

Sometimes just selecting an appropriate electrode material may not be enough and there is a need to coat or functionalize electrode material with chemicals so as to increase the electron transfer potential of that electrode. And with this objective, the surface of electrodes is modified with different chemicals and materials and may comprise several sets of treatments. Surface modifications like the use of carbon nanotube (CNT) powders on electrodes to increase biofilm activity and electrical conductivity have been in practice for almost a decade (Liang et al., 2011). Similarly, nitric acid treatment and heat treatment have also been reported for enhancement of electron transfer efficiency in the past (Dong et al., 2014). For ferrying of electrons from microbes to the electrodes, many redox mediators are also used, like 1,4-naphthoquinone (NQ), anthraquinone-1,6-disulfonic acid (AQDS), and neutral red (NR) (Lowy et al., 2006; Wang et al., 2011).

6.3.2   CATHODE MATERIAL

In EF, cathodes work as electron donors and it is here at this surface the reduction of the substrate to synthesize the desired product takes place. The materials about which we had discussed in the anode section earlier are also utilized for making the cathode more often. In practice, cathodes are mainly composed of packed and brush structures like carbon felt, granular graphite, and graphite fiber brush, mainly to ensure a large surface area for interactions (Wei et al., 2011).

However, for some cases, materials of the cathode are different from the anode, as catalysts are used in the cathode (Lovely & Holmes, 2021). The commonly used material for oxygen reduction is platinum-coated cathodes. Similarly, other noble metals like Pd, Mn, Zr, Ru, and Cu and metal oxides like ZnO, ZrO_2, etc. are used, subject to the desired product. Apart from these, to enhance electron transfer, even immobilization of enzyme and electron shuttles on the electrode have been successfully employed (refs). These modifications basically facilitate the spontaneous redox reactions towards high product recovery.

Srikanth et al. (2011) performed a comparative analysis of different electrode materials like graphite, aluminum, brass, copper, nickel, and stainless steel and concluded that this graphite showed better performance as an anode electrode. Graphite functions as good anode material with a high electron density (ED) and high microbial population (Srikanth et al., 2011). Nickel and stainless steel showed good anodic properties next to graphite as compared to brass, aluminium, and copper.

6.4 MICROBE-ELECTRODE INTERACTION

The microbe-electrode interactions in an electro-fermentation (EF) process are a key aspect of such fermentation processes or productions. The microorganism is a single-cell organism. During the metabolic pathway, it produces electrons or sometimes it requires an electron supply so it uses electrodes as electron donors or acceptors in the EF system (Choi & Sang, 2016).

Therefore, the microbes that can transport electrons over biological membranes are electroactive and they may further be categorized into two groups:

    i. Exoelectrogens
    ii. Endoelectrogens

Exoelectrogens are metal-reducing microorganisms that can transfer electrons to the electrode (anode). In contrast, endoelectrogens are metal-oxidizing microbes that can receive electrons from the electrode (cathode). Microbes play a central role in electro-fermentation by conducting bidirectional electron transfer (BET) to catalyzed reactions at the electrode (Logan, 2009; Gong et al., 2020) (Logan, 2009).

6.4.1 EXTRACELLULAR ELECTRON TRANSFER

Microorganisms can transfer electrons into and out of the cell from or towards an electron source or acceptor (i.e., electrodes) by a mechanism called extracellular electron transfer (EET). The EET mechanism works in two ways – i) direct EET and ii) indirect EET (Gardel & Girguis, 2013) (Figure 6.2). In direct EET, nanowires, cytochromes, and other redox proteins are involved, and in indirect EET, self-secreted molecules (e.g., phenazines, flavins, etc.) or external artificial electron shuttles (e.g., methyl viologen, neutral red, etc.) mediate the transfer of an electron (Rabaey & Rozendal, 2010; Gong et al., 2020).

6.4.1.1 Direct EET

Direct extracellular electron transfer takes place by physical contact between electroactive bacteria and electrodes. These electroactive bacteria do not require any external mediators and, instead, form a biofilm on the electrode's surface that promotes direct electron transfers. The direct transfer generally involves a series of periplasmic and outer-membrane complexes and also includes some

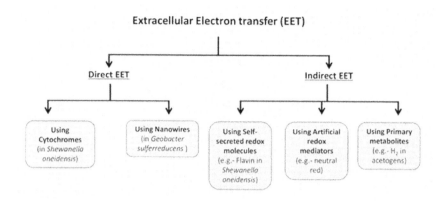

FIGURE 6.2 Types of extracellular electron transfer.

membrane-associated enzymes and exocellular appendages, like conductive pili or pilus-like structures (Gardel & Girguis, 2013; Lovely & Holmes, 2021) (Figure 6.3).

Shewanella oneidensis and *Geobacter sulferreducens* are two of the most extensively studied metal-reducing bacteria for extracellular electron transfer mechanisms (Edwards et al., 2020). In *Shewanella oneidensis*, *c*-type cytochrome is responsible for both electron uptake and delivery. Using *c*-type cytochrome *Cym*A, electrons are transported from the inner membrane to the outer membrane, where outer membrane proteins like *Mtr*A, *Mtr*B, and *Mtr*C

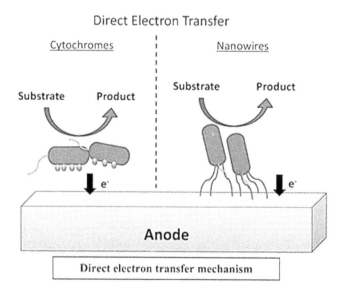

FIGURE 6.3 Setup for direct electron transfer.

collectively exchange the electrons with the exterior environment (i.e., electrode) using flavin as an electron shuttle or using nanowires (Figure 6.3) (Hartshorne et al., 2009; Edwards et al., 2020; Zheng et al., 2020). A nanowire is an extension of the outer membranes that connect electroactive bacteria with the electrode (Figure 6.3). In the last few years, the role of nanowires (pilus-like appendages) in electron transfer has been thoroughly established by several researchers (Reguera et al., 2005; Gardel & Girguis, 2013; Lovley & Holmes, 2021). In *Geobacter sulferreducens* also, the transfer of electrons takes place mainly by membrane-bound *c*-type cytochrome (*OmcE* and *OmcS*) and type IV pili (Holmes et al., 2006; Reguera et al., 2006; Chen et al., 2020).

However, recent research has proposed that even in the absence of cytochrome, other redox proteins like ferredoxin (*Rnf*) and rubredoxin (*Rub*) can facilitate the electron transfer mechanism (Kracke et al., 2015). In sulfate-reducing bacterial species like *Desulfovibrio* sp., rubredoxin was reported to function as a redox mediator. In bacteria belonging to Clostridia (*Clostridium ljungdahlii, Clostridium pasteurianum, and C. aceticum*) and in *Sporomusa ovate*, ferredoxin was found to serve as a redox mediator (Choi & Sang, 2016).

6.4.1.2 Indirect EET

Indirect EET is the shuttle-mediated electron transfer method in which the exchange of electrons takes place through the electron shuttle (redox mediators) (Figure 6.4). These electron shuttles are reduced or oxidized to transfer electrons from microbes to the electrode. Recent studies show the rate of electron transfer decreases if we eliminate mediators from the process. For example, removal of riboflavin from biofilms reduced the electron transfer rate by 70%, which further affects the product yield (Marsili et al., 2008; Barbir, 2009; Chen et al., 2020).

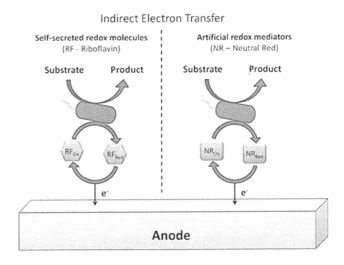

FIGURE 6.4 Setup for indirect electron transfer.

6.5   REDOX MEDIATORS

Redox mediators can be classified into three types: i) self-secreted redox molecules, ii) artificial redox mediators, and iii) primary metabolites (Gong et al., 2020, Chen et al., 2020).

6.5.1   SELF-SECRETED SMALL REDOX MOLECULES

Many microbes don't require external mediators and are capable of secreting their own redox mediators for electron transfer, such as flavins and riboflavin secreted by *Shewanella oneidensis*. Likewise, other self-secreted redox molecules have been discovered like phenazines synthesize by *Pseudomonas aeruginosa,* quinones by *Shewanella oenidensis*, and pyocyanin (PYO), 1-hydroxyphenazine (1-OHPHZ), phenazine-1-carboxylic acid (PCA), and phenazine-1-carboxamide (PCN) found in by *Pseudomonas* sp. (Kracke et al., 2015).

6.5.2   ARTIFICIAL REDOX MEDIATORS

Artificial redox mediators are the mediators that are added in the EF process externally. Methyl viologen, neutral red, potassium ferricyanide, anthraquinone-2, 6-disulfonate (AQDS), and artificial-synthesized phospholipid polymers can mediate the EET (Sund et al., 2007; Chen et al., 2020.

6.5.3   PRIMARY METABOLITES AS REDOX MEDIATORS

Many microorganisms use primary metabolites like H_2 and formate to transfer electrons. These primary metabolites are produced by microorganisms at electrodes. The use of H_2 as an electron donor in methanogens or acetogens have already been documented (Blanchet et al., 2015).

6.6   TYPES OF MICROBIAL CULTURES USED IN EF

There are two main types of cultures used in the electro-fermentation system: monoculture and mixed culture. Today, many researchers are working on mixed microbe cultures because they provide various advantages over monoculture. A single microorganism is used in monoculture, whereas in mixed culture, two or more than two microorganisms are used. Sometimes electron generating bacteria cannot transfer electrons to the electrode, so in that case, mixed cultures are helpful. In mixed culture, electroactive microorganisms are at work as electron mediators between electrodes and electron generating bacteria. Mixed culture may provide several advantages, such as removing the inhibitors and toxins from media, producing growth factors, and balancing the conditions required by other microorganisms (Kumar et al., 2017; San-Martín et al., 2019; Chen et al., 2020). An example of this is reported for *Desulfopila corrodens*, which accepts the electron from the cathode to generate hydrogen (H_2) as an intermediate;

thereafter, *Acetobacterium woodii* and *Methanococcus maripaludis* consume that H_2 to produce acetate and methane, respectively (Deutzmann et al., 2015; Deutzmann & Spormann, 2017; Logan et al., 2019; Yee et al., 2020). Mixed cultures aid in the overall electro-fermentation process, resulting in increased product yield.

6.7   COMMERCIAL PRODUCTS FROM THE EF PROCESS

6.7.1   VOLATILE FATTY ACIDS

Volatile fatty acids (VFA) are linear organic short-chain fatty acids with six or fewer carbon atoms, such as acetic acid, propionic acid, butyric acid, isobutyric acid, and isovaleric acid (Figure 6.5). VFAs have an industrial application in several sectors, such as pharmaceuticals, cosmetics, polymer, food and beverage, textiles, and plastic production (Baumann & Westermann, 2016). Also, VFAs can be used as precursors for the production of polyhydroxyalkanoate, generating electricity and biogas production (Mengmeng et al., 2009). Hence, VFAs are the major products to be recovered from electro-fermentation. Nowadays, researchers have begun to use sewage sludge for production of VFAs and are exploring all possibilities to further improve product yield (Ma et al., 2016). Volatile fatty acids (VFAs) derived from waste are generally acknowledged as a viable alternative to petroleum-based compounds. Mainly acetogenic and chain elongating microbes are involved in the conversion of organic and inorganic substrates to VFAs (Bhatia & Yang, 2017). The metabolic pathway followed by glucose to produce acids is mentioned in Figure 6.5 and a list of the different microbes and substrates used for VFA production are highlighted in Table. 6.1.

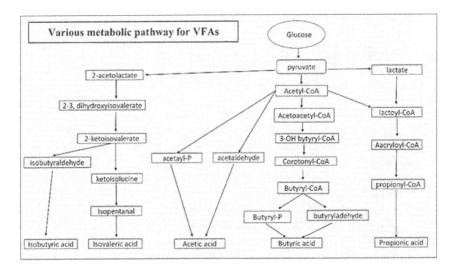

FIGURE 6.5   Scheme depicting the various metabolic pathways for volatile fatty acids.

TABLE 6.1
List of Different Microbes and Substrates Used for VFA Production

Product	Microorganism	Substrate	Concentration (gL^{-1})/yield (%)	Reference
Succinate	*Actinobacillus succinogenes*	Arabinose	4.7/31	(Zhao et al., 2016)
	Actinobacillus succinogenes	Glucose	7.88/53	(Zhao et al., 2016)
	Actinobacillus succinogenes	Xylose	5.24/35	(Zhao et al., 2016)
	Actinobacillus succinogenes	Corncob hydrolysate	3.84/26	(Zhao et al., 2016)
Acetate	Mixed culture	CO_2	95 mg d^{-1}	(Jiang et al., 2013)
	Enriched electroactive culture biofilm (mixed)	CO_2	2.35 mM d^{-1}	(Karthik et al., 2020)
	C. ljungdahlii DSM13528	CO_2	~105 µM	(Nevin et al., 2011)
	Moorella thermoacetica	CO_2	~90 µM	(Nevin et al., 2011)
Muconate	*C. glutamicum*	Catechol	85/100	(Bakonyi et al., 2018)
	E. coli sp.	Glucose	59/30	(Karthik et al., 2020)
	Pseudomonas DCB-71	Toluene	45/96	(Yoshikawa et al., 1990)
	Arthrobacter sp. T8626	Benzoate	44/96	(Karthik et al., 2020)

6.7.2 ACETIC ACID

Acetic acid can be a major end product as well as a platform chemical (serve as a precursor for many other syntheses). For acetate production, carbon dioxide (CO_2) and organic sugars (e.g., glucose) are the main substrate and, therefore, respective microbes that can reduce CO_2 effectively (like acetogens) are the best choice for acetate production (Marshall et al., 2013). *Clostridium ijungdahlii, Clostridium aceticum, Moorella thermoacetica, Sporomusa ovata, Sporomusa silvacetica,* and *Sporomusa sphaeroides* are some autotrophic bacteria that are testified as being able to produce acetate through electro-fermentation more effectively (Bajracharya et al., 2017).

6.7.2.1 Mechanism of Acetate Production

Initially, water is oxidized at the anode, which generates protons and electrons. Using applied potential, electrons transfer to the cathode while protons move through the membrane to the cathodic chamber. Further, these electrons are accepted by microbes at the cathode to reduce CO_2, wherein acetic acid gets synthesized as the product (Bajracharya et al., 2017).

The reactions that occur at electrodes are given below:

At anode,

$$4H_2O \quad \rightarrow \quad O_2 \quad + \quad 8H^+ \quad + \quad 8e^-$$

At cathode,

$$2HCO_3^- \quad + \quad 9H^+ \quad + \quad 8e^- \quad \rightarrow \quad CH_3COO^- \quad + \quad 3H_2O$$

Cell voltage $(E_{cell}) = E'_{cat} - E'_{anode} = -1.08$ V

The negative cell voltage indicates energy needs to be applied.

During reduction, CO_2 undergoes various metabolic reactions to produce acetate and are well explained by the Wood-Ljungdahl pathway (Figure 6.5) (Ragsdale & Pierce, 2008). Specific employment of biocatalysts through the use of a diversified mix of microbial cultures has been demonstrated to uplift acetate production as high as five times (Bajracharya et al., 2017). Bajracharya et al. (2017) describe how they achieved a maximum acetate production rate of 400 mg $L_{catholyte}^{-1}$ d^{-1} at -1 V (vs. Ag/AgCl) by removing methanogens from a mixed culture. This is because in the presence of methanogens, methane is formed and ultimately it decreases the yield of acetate. Also, the use of a granular graphite bed as the cathode can increase the yield of acetate, as granular graphite material has a larger surface area (Marshall et al., 2013; Zhou et al., 2019).

6.7.3 PROPIONIC ACID

Propionic acid is generally used as an anti-microbial agent and also has diverse applications in perfume, paint, and food industries. Propionic acid is also used as a food additive (Liu et al., 2012). Organic molecules such as glucose, xylose, lactose, and glycerol are the substrates used in the production of propionic acid. Most bacteria that produce propionic acid belong to the *Propionibacterium* species. Examples include *P. freudenreichii, P. acidipropionici, P. thoenii, and P. shermanii.* Liu et al. (Liu et al., 2016) developed an engineered *P. jenseniis* strain and was able to produce 34.93 gL^{-1} of propionic acid. They have deleted the lactate dehydrogenase (*ldh*) and pyruvate oxidase (*poxB*) gene, which helps to reduce by-products like lactate and acetate (Liu et al., 2016). They have also overexpressed the phosphoenolpyruvate carboxylase (*ppc*) gene, which avoids pyruvate intermediate and directly converts phosphoenolpyruvate to oxaloacetate.

6.7.4 BUTYRIC ACID

Butyric acid has diverse applications in food and pharmaceutical industries. Butyric acid is one of the major products of EF. Several microbes have been examined for their ability to produce butyric acid, such as *Butyrivibrio*, *Butyribacterium*, *Clostridium*, *Fusobacterium*, *Eubacterium*, *Megasphera*, and *Sarcina*.

Butyric acid could be produced in two ways (Figure 6.6):

  i. Directly from carbon dioxide through the Wood-Ljungdahl pathway by reduction of acetyl-CoA and
 ii. Using acetate, lactate, and ethanol through a chain-elongation pathway (reverse β-oxidation).

In this pathway, the first homoacetogenesis of CO_2 and H_2 takes place to form acetate. Then, ethanol is formed by acetate reduction. At the end, chain elongation of ethanol and acetate occurs to form butyrate. This pathway is carried out only in a specific environment (low glucose and limiting CO_2 concentration) (Figure 6.6; Raes et al., 2017; Batlle-Vilanova et al., 2017).

So far, a maximum titer of 5.5 g L^{-1} and production rates of up to 0.16 g L^{-1} d^{-1} have been reported. Different research groups have studied the role of mediators in butyric acid production (Batlle-Vilanova et al., 2017; Paiano et al., 2019). They reported that if we use mediators during the production of butyric acid, then only the *n*-isomer of butyric acid is produced, whereas, in the absence

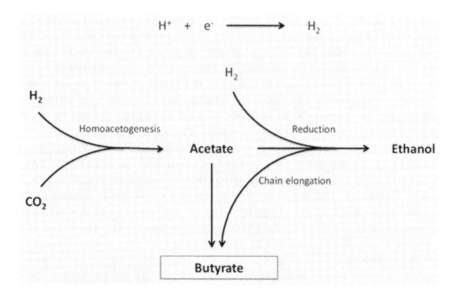

FIGURE 6.6   A diagrammatic display of a series of reactions taking place at the cathode to produce butyrate.

of mediators, both *i*- and *n*-isomers of butyric acid were formed. This indicates selectivity towards specific compounds can be achieved with the use of mediators. Currently, many researchers are exploring different ways to enhance the yield of butyric acid and some of them have been highlighted in Table 6.1.

6.7.5 POLY HYDROXYBUTYRATE

Poly (3-hydroxybutyrate) (P3HB) is produced by a variety of bacteria, mainly as a carbon reserve, and the industrial production by fermentation of glucose by the bacterium *Alcaligenes eutrophus*. P3HB is used in products ranging from simple plastic tableware to cutting-edge surgical stitches and surgical pins. A lot of P3HB research is also being done regarding its drug delivery potential (Gong et al., 2020).

Microbial electrosynthesis (MES) of CO_2 via assimilation in *Ralstonia eutropha* was performed by constructing a formate dehydrogenase (FDH)–assisted MES system. It helped to catalyze the reduction process of CO_2 to formate in the cathodic chamber. Formate helped in carrying electrons from the cathode to *R. eutropha*.

Formate helped in carrying electrons and also provided carbon for the synthesis of PHB. Eventually, the titer of PHB in genetically engineered *R. eutropha* was increased to 485 ± 13 mgL^{-1} (Chen et al., 2018).

6.7.6 ALCOHOL

For power generation, to some extent, we are relying on non-renewable resources (fossil fuels). Depletion of the oil supply and the adverse effect of fossil fuels on the atmosphere is a big concern today. Therefore, it is necessary to increase the production and use of other alternative biofuels like ethanol, butanol, and hydrogen for power generation. Electro-fermentation has the potential to be a key source of biofuel in the future. Several species of the *Enterobacteriaceae* family, such as *Citrobacter freundii* and *Klebsiella pneumonia,* are capable of fermenting glycerol to produce ethanol, butanol, and 1,3-propanediol (1, 3-PDO) (Yazdani & Gonzalez, 2007). Other than H_2 production in EF, alcohol can be produced by reduction of VFA production intermediate by the electrons generated at the anode. Speers et al. (2014) developed a microbial electrolysis cell (MEC) in which they used exoelectrogen *Geobacter sulfurreducens* and a bacterium, *Clostridium cellobioparum*, that produce ethanol in the anodic chamber by glycerol fermentation, whereas in the cathodic chamber hydrogen was generated (Figure 6.7a and b).

6.7.6.1 Butanol

Butanol is a four-carbon alcohol that can be used as a solvent for paint and has various applications in the textile industry. It is a better alternative for many non-renewable resources as it could be used as a biofuel. Glucose and glycerol are the major substrates used for butanol production in the electro-fermentation system.

(a)

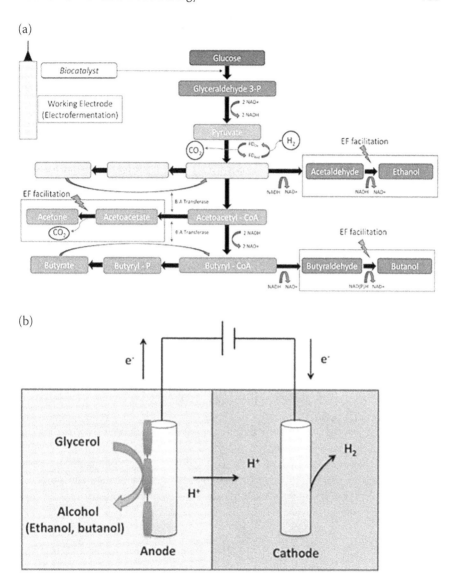

(b)

FIGURE 6.7 (a) Scheme showing metabolic pathway followed by glucose to produce alcohol; (b) anodic chamber setup for alcohol production by glycerol fermentation.

During fermentation, the metabolic pathway followed by glucose and glycerol is mentioned in Figure 6.7b. Various microorganisms produce butanol naturally. The most important strains belong to the genus *Clostridium*: *C. acetobutylicum, C. beijerinckii, C. saccharobutylicum*, and *C. saccharoperbutylacetonicum*. There are several other microorganisms that produce butanol: *Butyribacterium* (*B. methylotrophicum*) and *Thermoanaerobacterium* (*T. thermosaccharolyticum*)

(Schiel-Bengelsdorf et al., 2013). Khosravanipour Mostafazadeh et al. (2016) achieved a yield as high as 13.31 gL^{-1} by using glucose as a substrate and *C. pasteurianum* as a biocatalyst at 1.32 V of applied potential.

6.7.7 1,3-PROPANEDIOL (1, 3-PDO)

The 1,3-PDO is a value-added industrial solvent that can be produced by performing glycerol fermentation. Kim et al. (2020) performed cathode electrode-driven fermentation to produce 1,3-PDO from glycerol with several electron shuttles (2-hydroxy-1,4-naphthoquinone, neutral red, and hydroquinone) using the *Klebsiella pneumoniae L17* strain. They observed that homologous overexpression of *Dha*B and *Dha*T enzymes in *Klebsiella pneumoniae L17* enhanced the 1,3-PDO production. Electro-fermentation showed an increased 1,3-PDO productivity by as much as 32% when compared to anaerobic fermentation (Moscoviz et al., 2018). Researchers even used food waste as a substrate for the production of alcohols and other value-added products by the EF strategy (Huang et al., 2015).

6.7.8 HYDROGEN

Climate change has become a serious plight. Green alternative fuel is on demand to replace fossil fuel, which is a major contributor to pollution. Hydrogen acts as an ideal fuel as it is environmentally friendly. In the present scenario, hydrogen gas is produced by water electrolysis, biogas, etc., which is not at all feasible for long-term purposes. That's where hydrogen production via electro-fermentation becomes a crucial technology.

Hydrogen can be produced via dark fermentation. Theoretically, if we look at the stoichiometry, we can see that it will yield a maximum of 4 mol of hydrogen per mole of glucose fermented (H. Liu et al., 2005), with acetate and 2 mol/mol glucose as end products. However, it was found that integrating such a process with anaerobic MEC will enhance the hydrogen production by fourfold. The dark fermentation reaction:

$$C_6H_{12}O_6 \; + \; 2H_2O \; \rightarrow \; 4H_2 \; + \; 2CO_2 \; + \; 2C_2H_4O_2$$

$$C_6H_{12}O_6 \; \rightarrow \; 2H_2 \; + \; 2CO_2 \; + \; C_4H_8O_2$$

With butyrate and acetate, the achievable hydrogen yield is only 2 mol/mol of glucose and 4 mol/mol of glucose, respectively, but integrating this with MEC using bacteria such as *Pseudomonas geobacter, Schwanella,* etc., leads to the production of H_2 in anodes and cathodes by these microbes. The reaction:

Anode reaction: $C_2H_4O_2 \; + \; 2H_2O \; \rightarrow 2CO_2 \; + 8e^- \; + 8H^+$

Cathode reaction: $8H^+ + 8e^- \rightarrow 4H_2$

The effluent produced by hydrogen-producing bioreactors are intermediate metabolites that can act as an alternative to the synthetic substrate for hydrogen-producing MECs, which significantly improve the efficiency of MEC and increases the yield (Venkata Mohan et al., 2016); thus, making it economically viable. Hydrogen production by MEC (250 mV) is more energy efficient compared to electrolysis (1.23 V). Also, as the process is carried out in anaerobic conditions, the Coulombic efficiency increases. As a result, an utmost yield of 11 mol of hydrogen/mol of glucose can be obtained at the rate of 1 m^3hydrogen/day/m^3 in the reactor. It was seen that struvite ($MgNH_4PO_4 \cdot 6H_2O$) crystallization occurs at the cathode during a bioelectochemical reaction in a single-chamber MEC. Struvite is a phosphate fertilizer that can be used in agriculture. Production of hydrogen can be enhanced by eliminating membranes from dual chambers of MEC. However, this might lead to the utilization of hydrogen by methanogen to make methane.

6.7.9 METHANE

Methane is an important molecule that is primarily used as biofuel to make heat and electricity, and it is also a precursor of various chemicals. Methane is economical and has diverse applications in industries. Methane could be considered a powerhouse of energy (Hwang et al., 2018). Methanogenesis is the process by which methanogens produce methane by reducing carbon dioxide. Organic molecules like formate, acetate, and methylamine are used with CO_2 and H_2 as substrates for methane production. Even though the substrates are very simple, methane formation is a complex biochemical process that involves various coenzymes and genes. Methanogens are categorized into three groups according to the substrate used: acetolactic methanogens, hydrogenotrophic methanogens, and methylotrophic methanogens (Table 6.2) (Fu et al., 2021).

During the EF process, microbial oxidation of the organic matter takes place, releasing both CO_2 and electrons at the anode. This CO_2 is used as a substrate in the cathodic chamber and then electrons are further transferred from anode to cathode. Later, electrons are accepted by methanogens present at the cathode and reduce CO_2 to form methane (CH_4). One mole of CO_2 is converted to 1 mol of methane as a result of autotrophic methanogenesis (Mayer et al., 2019).
Overall reaction:

$$CO_2 + 8H^+ + 8e^- \rightarrow CH_4 + 2H_2O$$

M. maripaludis is very efficient when it comes to methane production by microbial electrosynthesis with the productivity of (8.81 ± 0.51 mmol m^{-2} d^{-1}) and the Coulombic efficiency ($58.9 \pm 0.8\%$) (Mayer et al., 2019). Methanogenesis can be improved by stimulating electron transfer from cathode to methanogens. The efficiency of methane production can be determined by electron exchange

TABLE 6.2
Methanogens Used for Methane Production According to Substrate

Methanogens	Substrate	Reactions	Typical methanogens
Hydrogenotrophic methanogens	H_2 and CO_2 Formate Methanol	$4H_2 + CO_2 \rightarrow CH_4 + 2H_2O$ $4HCOOH \rightarrow CH_4 + 3CO_2 +$ $2H_2O$ $4CH_3OH \rightarrow 3CH_4 + CO_2 +$ $2H_2O$	*Methanobacterium bryantii* *Methanobacterium formicicum* *Methanobacterium* *thermoalcaliphium* *Methanothermobacter* *thermoautotrophicum* *Methanothermobacter wolfeii* *Methanobrevibacter smithii* *Methanobrevibacter* *ruminantium* *Methanococcoides* *methylutens*
Aceticlastic methanogens	Acetate	$CH_3COOH \rightarrow CH_4 + CO_2$	*Methanosaeta concilii* *(soehngenii)* *Methanosaeta thermophila*
Methylotrophic methanogens	Trimethylamine Dimethyl sulfate Methylated ethanolamines	$4(CH3)_3N + 6H_2O \rightarrow 9CH_4$ $+3CO_2+4NH_3$ $2(CH3)_2S + 2H_2O \rightarrow$ $3CH_4+CO_2 + H_2S2$ $(CH3)_2NH + 2H_2O \rightarrow$ $3CH_4+CO_2 + 2NH_3$	*Methanosarcina barkeri* *Methanosarcina mazei* *Methanosarcina thermophile*

on the cathode (Beckmann et al., 2016); it was demonstrated that the addition of the mediator neutral red increases the efficiency of electron transfer without an applied potential, which leads to an increase in methane production.

6.8 SUMMARY AND CONCLUSIONS

The process of electro-fermentation has improved the efficiency of substrate conversion remarkably. The recent advancement in this field has enabled us to generate electricity using waste from various sources. The process of microbial electrosynthesis has proved to be an alternative option for the economical mass production of useful chemicals from pure substrates as well as biowastes. Because of the two separate chambers, waste organic matter existing in the anodic chamber does not affect the synthesis of product in the cathodic chamber. EF, because of its several advantages, such as low cost, better efficiency, and solution for several problems faced in the conventional fermentation process, is seen as a potential key process. The future of this technology is broad, and it provides huge opportunities through various applications and integrations with other technologies, thereby stimulating the development of the industry and has a very high capability of being a sustainable approach for the mass production of fuels and other chemicals. However, this technology has some drawbacks that can be resolved with further research and development in this area.

6.9 FUTURE PROSPECTS

More intensive research is required to understand the microbial-electrode interactions and also to know about the exoelectrogenic activity to improve the electron transfer mechanism. It is necessary to specifically develop biocompatible electrodes that are economical, flexible, and catalytically active characteristics to achieve efficient electron kinetic. To enhance the yield of the product, we should try to inhibit side products that can also reduce the separation cost. A metabolic engineering approach to develop highly efficient electroactive microbe systems can extend utilities of the EF system significantly in the near future. However, we still need to opt for a multidisciplinary approach in this field of research.

REFERENCES

Bajracharya, S., Srikanth, S., Mohanakrishna, G., Zacharia, R., Strik, D. P., & Pant, D. (2017). Biotransformation of carbon dioxide in bioelectrochemical systems: State of the art and future prospects. *Journal of Power Sources, 356*, 256–273. https://doi.org/10.1016/j.jpowsour.2017.04.024

Bakonyi, P., Kumar, G., Koók, L., Tóth, G., et al. (2018). Microbial electrohydrogenesis linked to dark fermentation as integrated application for enhanced biohydrogen production: A review on process characteristics, experiences and lessons. *Bioresource Technology, 251*, 381–389.

Barbir, F. (2009). Fuel cells – Exploratory fuel cells | Regenerative fuel cells. In: *Encyclopedia of Electrochemical Power Sources*, 224–237. https://doi.org/10.1016/B978-044452745-5.00288-4

Batlle-Vilanova, P., Ganigué, R., Ramió-Pujol, S., Bañeras, L., Jiménez, G., Hidalgo, M., Balaguer, M. D., Colprim, J., & Puig, S. (2017). Microbial electrosynthesis of butyrate from carbon dioxide: Production and extraction. *Bioelectrochemistry, 117*, 57–64. https://doi.org/10.1016/j.bioelechem.2017.06.004

Baumann, I., & Westermann, P. (2016). Microbial production of short chain fatty acids from lignocellulosic biomass: Current processes and market. *BioMed Research International, 2016*, 8469357. https://doi.org/10.1155/2016/8469357

Beckmann, S., Welte, C., Li, X., Oo, Y. M., Kroeninger, L., Heo, Y., Zhang, M., Ribeiro, D., Lee, M., Bhadbhade, M., Marjo, C. E., Seidel, J., Deppenmeier, U., & Manefield, M. (2016). Novel phenazine crystals enable direct electron transfer to methanogens in anaerobic digestion by redox potential modulation. *Energy and Environmental Science, 9*(2), 644–655. https://doi.org/10.1039/c5ee03085d

Bhagchandanii, D. D., Babu, R. P., Sonawane, J. M., Khanna, N., et al. (2020). A comprehensive understanding of electro-fermentation. *Fermentation, 6*(3), 92.

Bhatia, S. K., & Yang, Y. H. (2017). Microbial production of volatile fatty acids: Current status and future perspectives. *Reviews in Environmental Science and Biotechnology, 16*(2), 327–345. https://doi.org/10.1007/s11157-017-9431-4

Blanchet, E., Duquenne, F., Rafrafi, Y., Etcheverry, L., Erable, B., & Bergel, A. (2015). Importance of the hydrogen route in up-scaling electrosynthesis for microbial CO_2 reduction. *Energy and Environmental Science*, *8*(12), 3731–3744. https://doi.org/ 10.1039/c5ee03088a

Bursac, T., Gralnick, J. A., & Gescher, J. (2017). Acetoin production via unbalanced fermentation in *Shewanella oneidensis*. *Biotechnology and Bioengineering*, *114*(6), 1283–1289. https://doi.org/10.1002/bit.26243

Chandrasekhar, K., Amulya, K., & Venkata Mohan, S. (2014). Solid phase bio-electrofermentation of food waste to harvest value-added products associated with waste remediation. *Waste Management*, *45*, 57–65.

Chandrasekhar, K., Kumar, A. N., Raj, T., Kumar, G., & Kim, S.-H. (2021a). Bioelectrochemical system-mediated waste valorization. *Systems Microbiology and Biomanufacturing*, *1*, 1–12. https://doi.org/10.1007/S43393-021-00039-7

Chandrasekhar, K., Lee, Y. J., & Lee, D. W. (2015). Biohydrogen production: Strategies to improve process efficiency through microbial routes. *International Journal of Molecular Sciences*, *16*, 8266–8293.

Chandrasekhar, K., Mehrez, I., Kumar, G., & Kim, S.-H. (2021b). Relative evaluation of acid, alkali, and hydrothermal pretreatment influence on biochemical methane potential of date biomass. *Journal of Environmental Chemical Engineering*, *9*, 106031. https://doi.org/10.1016/J.JECE.2021.106031

Chandrasekhar, K., Naresh Kumar, A., Kumar, G., Kim, D. H., Song, Y. C., & Kim, S. H. (2021c). Electro-fermentation for biofuels and biochemicals production: Current status and future directions. *Bioresource Technology*, *323*, 124598.

Chandrasekhar, K. & Venkata Mohan, S. (2012). Bio-electrochemical remediation of real field petroleum sludge as an electron donor with simultaneous power generation facilitates biotransformation of PAH: Effect of substrate concentration. *Bioresource Technology*, 110, 517–525.

Chen H., Dong F., & Minteer S. D. (2020). The progress and outlook of bioelectrocatalysis for the production of chemicals, fuels and materials. *Nature Catalysis*, *3*(3), 225–244. https://doi.org/10.1038/s41929-019-0408-2

Chen, X., Cao, Y., Li, F., Tian, Y., & Song, H. (2018). Enzyme-assisted microbial electrosynthesis of poly(3-hydroxybutyrate) via CO_2 bioreduction by engineered *Ralstonia eutropha*. *ACS Catalysis*, *8*(5), 4429–4437.

Choi, O., & Sang, B. I. (2016). Extracellular electron transfer from cathode to microbes: Application for biofuel production. *Biotechnology for Biofuels*, *9*(1), 1–14.

Cok, B., et al. (2014). Succinic acid production derived from carbohydrates: An energy and greenhouse gas assessment of a platform chemical toward a bio-based economy. *Biofuels, Bioproducts and Biorefining*, *8*, 16–29.

Deutzmann, J. S., Sahin, M., & Spormann, A. M. (2015). Extracellular enzymes facilitate electron uptake in biocorrosion and bioelectrosynthesis. *mBio*, *6*(2), 1–8. https:// doi.org/10.1128/mBio.00496-15

Deutzmann, J. S., & Spormann, A. M. (2017). Enhanced microbial electrosynthesis by using defined co-cultures. *ISME Journal*, *11*(3), 704–714.

Dong, Y., Zhou, Y., Ding, Y., Chu, X., & Wang, C. (2014). Sensitive detection of Pb(ii) at gold nanoparticle/polyaniline/graphene modified electrode using differential pulse anodic stripping voltammetry. *Analytical Methods*, *6*(23), 9367–9374.

Edwards, M. J., White, G. F., Butt, J. N., Richardson, D. J., & Clarke, T. A. (2020). The crystal structure of a biological insulated transmembrane molecular wire. *Cell*, *181*(3), 665–673.e10.

Enamala, M. K., Pasumarthy, D. S., Gandrapu, P. K., Chavali, M., Mudumbai, H., & Kuppam, C. (2019). Production of a variety of industrially significant products by biological sources through fermentation. In: P. K. Arora (Ed.), *Microbial Technology for the Welfare of Society* (pp. 201–221). Springer Singapore, Singapore. https://doi.org/10.1007/978-981-13-8844-6_9

Endreny, T., Avignone-Rossa, C., & Nastro, R. A. (2020). Generating electricity with urban green infrastructure microbial fuel cells. *Journal of Cleaner Production, 263,* 121337. https://doi.org/10.1016/J.JCLEPRO.2020.121337

Fu, S., Angelidaki, I., & Zhang, Y. (2021). In situ biogas upgrading by CO_2-to-CH_4 bioconversion. *Trends in Biotechnology, 39*(4), 336–347.

Gardel, E. J., & Girguis, P. R. (2013). *Microbe-electrode interactions: The chemico-physical environment and electron transfer, 3600169,* 156.

Gong, Z., Yu, H., Zhang, J., Li, F., & Song, H. (2020). Microbial electro-fermentation for synthesis of chemicals and biofuels driven by bi-directional extracellular electron transfer. *Synthetic and Systems Biotechnology, 5*(4), 304–313.

Hartshorne, R. S., Reardon, C. L., Ross, D., Nuester, J., Clarke, T. A., Gates, A. J., Mills, P. C., Fredrickson, J. K., Zachara, J. M., Shi, L., Beliaev, A. S., Marshall, M. J., Tien, M., Brantley, S., Butt, J. N., & Richardson, D. J. (2009). Characterization of an electron conduit between bacteria and the extracellular environment. *Proceedings of the National Academy of Sciences of the United States of America, 106*(52), 22169–22174.

Holmes, D. E., Chaudhuri, S. K., Nevin, K. P., Mehta, T., Methé, B. A., Liu, A., Ward, J. E., Woodard, T. L., Webster, J., & Lovley, D. R. (2006). Microarray and genetic analysis of electron transfer to electrodes in *Geobacter sulfurreducens. Environmental Microbiology, 8*(10), 1805–1815.

Huang, H., Singh, V., & Qureshi, N. (2015). Butanol production from food waste: A novel process for producing sustainable energy and reducing environmental pollution. *Biotechnology for Biofuels, 8*(1), 1–12.

Hwang, I. Y., Nguyen, A. D., Nguyen, T. T., Nguyen, L. T., Lee, O. K., & Lee, E. Y. (2018). Biological conversion of methane to chemicals and fuels: Technical challenges and issues. *Applied Microbiology and Biotechnology, 102*(7), 3071–3080.

Kadier, A., Chandrasekhar, K., & Kalil, M. S. (2017). Selection of the best barrier solutions for liquid displacement gas collecting metre to prevent gas solubility in microbial electrolysis cells. *International Journal of Renewable Energy Technology, 8,* 93. https://doi.org/10.1504/IJRET.2017.086807

Karthik, O., Reddy, C. N., Mehariya, S., & Banu, R. J. (2020). *Electro-Fermentation of biomass for high-value organic acids.* Metadata of the chapter that will be visualized online. October.

Khosravanipour Mostafazadeh, A., Drogui, P., Brar, S. K., Tyagi, R. D., Le Bihan, Y., Buelna, G., & Rasolomanana, S. D. (2016). Enhancement of biobutanol production by electromicrobial glucose conversion in a dual chamber fermentation cell using *C. pasteurianum. Energy Conversion and Management, 130,* 165–175.

Kim, C., Lee, J. H., Baek, J., Kong, D. S., Na, J. G., Lee, J., Sundstrom, E., Park, S., & Kim, J. R. (2020). Small current but highly productive synthesis of 1,3-propanediol from glycerol by an electrode-driven metabolic shift in *Klebsiella pneumoniae* L17. *ChemSusChem, 13*(3), 564–573.

Kracke, F., Lai, B., Yu, S., & Krömer, J. O. (2018). Balancing cellular redox metabolism in microbial electrosynthesis and electro fermentation – A chance for metabolic engineering. *Metabolic Engineering, 45,* 109–120.

Kracke, F., Vassilev, I., & Krömer, J. O. (2015). Microbial electron transport and energy conservation – The foundation for optimizing bioelectrochemical systems. *Frontiers in Microbiology*, *6*(June), 1–18.

Kumar, A., Hsu, L. H. H., Kavanagh, P., Barrière, F., et al. (2017). The ins and outs of microorganism–electrode electron transfer reactions. *Nature Reviews in Chemistry*, *1*, 1–13.

Kumar, P., Chandrasekhar, K., Kumari, A., Sathiyamoorthi, E., & Kim, B. S. (2018). Electro-fermentation in aid of bioenergy and biopolymers. *Energies*, *11*(2), 343.

Lee, C. S., Aroua, M. K., Daud, W. M. A. W., Cognet, P., Peres-Lucchese, Y., Fabre, P. L., et al. (2015). A review: Conversion of bioglycerol into 1,3-propanediol via biological and chemical method. *Renewable and Sustainable Energy Reviews*, *42*, 963–972.

Lee, H. S., Dhar, B. R., An, J., Rittmann, B. E., et al. (2016). The roles of biofilm conductivity and donor substrate kinetics in a mixed-culture biofilm anode. *Environmental Science and Technology*, *50*, 12799–12807.

Lee, J. W., Kim, H. U., Choi, S., Yi, J., et al (2011). Microbial production of building block chemicals and polymers. *Current Opinion in Biotechnology*, *22*(6), 758–767.

Lee, J. Y., Lee, S. H., & Park, H. D. (2016). Enrichment of specific electro-active microorganisms and enhancement of methane production by adding granular activated carbon in anaerobic reactors. *Bioresource Technology*, *205*, 205–212.

Liang, P., Wang, H., Xia, X., Huang, X., Mo, Y., Cao, X., & Fan, M. (2011). Carbon nanotube powders as electrode modifier to enhance the activity of anodic biofilm in microbial fuel cells. *Biosensors and Bioelectronics*, *26*(6), 3000–3004.

Liu, H., Grot, S., & Logan, B. E. (2005). Electrochemically assisted microbial production of hydrogen from acetate. *Environmental Science and Technology*, *39*(11), 4317–4320.

Liu, L., Guan, N., Zhu, G., Li, J., Shin, H. D., Du, G., & Chen, J. (2016). Pathway engineering of *Propionibacterium jensenii* for improved production of propionic acid. *Scientific Reports*, *6*(September 2015), 1–9.

Liu, L., Zhu, Y., Li, J., Wang, M., Lee, P., Du, G., & Chen, J. (2012). Microbial production of propionic acid from propionibacteria: Current state, challenges and perspectives. *Critical Reviews in Biotechnology*, *32*(4), 374–381.

Liu, X., Wang, S., Xu, A., Zhang, L. et al. (2019). Biological synthesis of high-conductive pili in aerobic bacterium *Pseudomonas aeruginosa*. *Applied Microbiology and Biotechnology*. *103*, 1535–1544.

Logan, B. E. (2009). Exoelectrogenic bacteria that power microbial fuel cells. *Nature Reviews Microbiology*, *7*(5), 375–381.

Logan, B. E., Rossi, R., Ragab, A., & Saikaly, P. E. (2019). Electroactive microorganisms in bioelectrochemical systems. *Nature Reviews in Microbiology*, *17*, 307–319.

Lovley, D. R., & Holmes, D. E. (2021). Electromicrobiology: the ecophysiology of phylogenetically diverse electroactive microorganisms. *Nature Reviews in Microbiology*, 20, 5–19. https://doi.org/10.1038/s41579-021-00597-6

Lowy, D. A., Tender, L. M., Zeikus, J. G., Park, D. H., & Lovley, D. R. (2006). Harvesting energy from the marine sediment-water interface II. Kinetic activity of anode materials. *Biosensors and Bioelectronics*, *21*(11), 2058–2063.

Ma, H., Chen, X., Liu, H., Liu, H., & Fu, B. (2016). Improved volatile fatty acids anaerobic production from waste activated sludge by pH regulation: Alkaline or neutral pH? *Waste Management*, *48*, 397–403.

Marshall, C. W., Ross, D. E., Fichot, E. B., Norman, R. S., & May, H. D. (2013). Long-term operation of microbial electrosynthesis systems improves acetate production by autotrophic microbiomes. *Environmental Science and Technology*, *47*(11), 6023–6029.

Marsili, E., Baron, D. B., Shikhare, I. D., Coursolle, D., Gralnick, J. A., & Bond, D. R. (2008). Shewanella secretes flavins that mediate extracellular electron transfer. *Proceedings of the National Academy of Sciences of the United States of America*, *105*(10), 3968–3973.

Mayer, F., Enzmann, F., Lopez, A. M., & Holtmann, D. (2019). Performance of different methanogenic species for the microbial electrosynthesis of methane from carbon dioxide. *Bioresource Technology*, *289*(June), 121706.

Mengmeng, C., Hong, C., Qingliang, Z., Shirley, S. N., & Jie, R. (2009). Optimal production of polyhydroxyalkanoates (PHA) in activated sludge fed by volatile fatty acids (VFAs) generated from alkaline excess sludge fermentation. *Bioresource Technology*, *100*(3), 1399–1405.

Moscoviz, R., Toledo-Alarcón, J., Trably, E., & Bernet, N. (2016). Electro-Fermentation: How to drive fermentation using electrochemical systems. *Trends in Biotechnology*, *34*(11), 856–865.

Moscoviz, R., Trably, E., & Bernet, N. (2018). Electro-fermentation triggering population selection in mixed-culture glycerol fermentation. *Microbial Biotechnology*, *11*(1), 74–83.

Nastro, R. A. (2014). Microbial fuel cells in waste treatment: Recent advances. *International Journal of Performability Engineering*, *10*, 367. https://doi.org/10. 23940/IJPE.14.4.P367.MAG

Nastro, R. A., Suglia, A., Pasquale, V., Toscanesi, M., Trifuoggi, M., & Guida, M. (2014). Efficiency measures of polycyclic aromatic hydrocarbons bioremediation process through ecotoxicological tests. *International Journal of Performability Engineering*, *10*, 411–418.

Paiano, P., Menini, M., Zeppilli, M., Majone, M., & Villano, M. (2019). Electro-fermentation and redox mediators enhance glucose conversion into butyric acid with mixed microbial cultures. *Bioelectrochemistry*, *130*, 107333.

Rabaey, K., & Rozendal, R. A. (2010). Microbial electrosynthesis – Revisiting the electrical route for microbial production. *Nature Reviews Microbiology*, *8*(10), 706–716.

Raes, S. M. T., Jourdin, L., Buisman, C. J. N., & Strik, D. P. B. T. B. (2017). Continuous long-term bioelectrochemical chain elongation to butyrate. *ChemElectroChem*, *4*(2), 386–395.

Ragsdale, S. W., & Pierce, E. (2008). Acetogenesis and the Wood-Ljungdahl pathway of CO_2 fixation. *Biochimica et Biophysica Acta – Proteins and Proteomics*, *1784*(12), 1873–1898.

Raj, T., Chandrasekhar, K., Banu, R., Yoon, J.-J., Kumar, G., & Kim, S.-H. (2021a). Synthesis of γ-valerolactone (GVL) and their applications for lignocellulosic deconstruction for sustainable green biorefineries. *Fuel*, *303*, 121333. https://doi.org/ 10.1016/J.FUEL.2021.121333

Raj, T., Chandrasekhar, K., Kumar, A. N., & Kim, S.-H. (2021b). Recent biotechnological trends in lactic acid bacterial fermentation for food processing industries. *Systems Microbiology and Biomanufacturing*, *2021*, 1–27. https://doi.org/10.1007/S43393-021-00044-W

Reguera, G., McCarthy, K. D., Mehta, T., Nicoll, J. S., Tuominen, M. T., & Lovley, D. R. (2005). Extracellular electron transfer via microbial nanowires. *Nature*, *435*(7045), 1098–1101. https://doi.org/10.1038/nature03661

Reguera, G., Nevin, K. P., Nicoll, J. S., Covalla, S. F., Woodard, T. L., & Lovley, D. R. (2006). Biofilm and nanowire production leads to increased current in Geobacter sulfurreducens fuel cells. *Applied and Environmental Microbiology*, *72*(11), 7345–7348. https://doi.org/10.1128/AEM.01444-06

Richter, H., McCarthy, K., Nevin, K. P., Johnson, J. P., Rotello, V. M., & Lovley, D. R. (2008). Electricity generation by *Geobacter sulfurreducens* attached to gold electrodes. *Langmuir, 24*(8), 4376–4379.

Roy, S., Schievano, A., & Pan, D. (2016). Electro-stimulated microbial factory for value added product synthesis. *Bioresource Technology. 213*, 129–139.

San-Martín, M. I., Sotres, A., Alonso, R. M., Díaz-Marcos, J., et al. (2019). Assessing anodic microbial populations and membrane ageing in a pilot microbial electrolysis cell. *International Journal of Hydrogen Energy, 44*, 17304–17315.

Santoro, C., Arbizzani, C., Erable, B., & Ieropoulos, I. (2017). Microbial fuel cells: From fundamentals to applications. A review. *Journal of Power Sources, 356*, 225–244.

Schiel-Bengelsdorf, B., Montoya, J., Linder, S., & Dürre, P. (2013). Butanol fermentation. *Environmental Technology (United Kingdom), 34*(13–14), 1691–1710. https://doi.org/10.1080/09593330.2013.827746

Schievano, A., Pepé Sciarria, T., Vanbroekhoven, K., De Wever, H., Puig, S., Andersen, S. J., Rabaey, K., & Pant, D. (2016). Electro-Fermentation – Merging electrochemistry with fermentation in industrial applications. *Trends in Biotechnology, 34*(11), 866–878. https://doi.org/10.1016/j.tibtech.2016.04.007

Sindhu, R. et al. (2011). Production and characterization of poly-3-hydroxybutyrate from crude glycerol by *Bacillus sphaericus* NII 0838 and improving its thermal properties by blending with other polymers. *Brazilian Archives in Biological Technology, 54*, 783–794.

Soccol, C. R., Pandey, A., Larroche, C., Pandey, A., & Larroche, C. (2013). In *Fermentation processes engineering in the food industry*. CRC Press, Boca Raton, FL, USA. ISBN 978-0-429-11200-3.

Speers, A. M., Young, J. M., & Reguera, G. (2014). Fermentation of glycerol into ethanol in a microbial electrolysis cell driven by a customized consortium. *Environmental Science and Technology, 48*(11), 6350–6358. https://doi.org/10.1021/es500690a

Srikanth, S., Pavani, T., Sarma, P. N., & Venkata Mohan, S. (2011). Synergistic interaction of biocatalyst with bio-anode as a function of electrode materials. *International Journal of Hydrogen Energy, 36*(3), 2271–2280. https://doi.org/10.1016/j.ijhydene.2010.11.031

Sund, C. J., McMasters, S., Crittenden, S. R., Harrell, L. E., & Sumner, J. J. (2007). Effect of electron mediators on current generation and fermentation in a microbial fuel cell. *Applied Microbiology and Biotechnology, 76*(3), 561–568. https://doi.org/10.1007/s00253-007-1038-1

Velvizhi, G., & Venkata Mohan, S. (2015). Bioelectrogenic role of anoxic microbial anode in the treatment of chemical wastewater: Microbial dynamics with bioelectro-characterization. *Water Research, 70*, 52–63. https://doi.org/10.1016/j.watres.2014.11.002

Venkata Mohan, S. & Chandrasekhar, K. (2011a). Self-induced bio-potential and graphite electron accepting conditions enhances petroleum sludge degradation in bio-electrochemical system with simultaneous power generation. *Bioresource Technology, 102*, 9532–9541.

Venkata Mohan, S. & Chandrasekhar, K. (2011b). Solid phase microbial fuel cell (SMFC) for harnessing bioelectricity from composite food waste fermentation: Influence of electrode assembly and buffering capacity. *Bioresource Technology, 102*, 7077–7085

Venkata Mohan, S., Chandrasekhar, K., Chiranjeevi, P., & Babu, P. S. (2013). Chapter 10 – Biohydrogen production from wastewater A2 – Pandey, Ashok. In: J.-S. Chang, P. C. Hallenbecka, C. Larroche (Eds.), *Biohydrogen* (pp. 223–257). Elsevier, Amsterdam. https://doi.org/10.1016/B978-0-444-59555-3.00010-6

Venkata Mohan, S., Nikhil, G. N., Chiranjeevi, P., Nagendranatha Reddy, C., Rohit, M. V., Kumar, A. N., & Sarkar, O. (2016). Waste biorefinery models towards sustainable circular bioeconomy: Critical review and future perspectives. *Bioresource Technology, 215,* 2–12. https://doi.org/10.1016/j.biortech.2016.03.130

Venkata Mohan, S., Prathima Devi, M., Venkateswar Reddy, M., Chandrasekhar, K., Asha Juwarkar, & Sarma, P. N. (2011). Bioremediation of real field petroleum sludge by mixed consortia under anaerobic conditions: Influence of biostimulation and bioaugmentation. *Environmental Engineering and Management Journal, 10*(11), 1609–1616.

Wang, K., Liu, Y., & Chen, S. (2011). Improved microbial electrocatalysis with neutral red immobilized electrode. *Journal of Power Sources, 196*(1), 164–168. https://doi.org/10.1016/j.jpowsour.2010.06.056

Wei, J., Liang, P., & Huang, X. (2011). Recent progress in electrodes for microbial fuel cells. *Bioresource Technology, 102*(20), 9335–9344. https://doi.org/10.1016/j.biortech.2011.07.019

Yazdani, S. S., & Gonzalez, R. (2007). Anaerobic fermentation of glycerol: A path to economic viability for the biofuels industry. *Current Opinion in Biotechnology, 18*(3), 213–219.

Yee, M. O., Deutzmann, J. S., Spormann, A. M. & Rotaru, A. E. (2020). Cultivating electroactive microbes – From field to bench. *Nanotechnology, 31,* 174003.

Zheng, S., Liu, F., Wang, B., Zhang, Y. et al. (2020). A methanobacterium capable of direct interspecies electron transfer. *Environmental Science and Technology, 54,* 15347–15354.

Zhou, M., Yan, B., Lang, Q., & Zhang, Y. (2019). Elevated volatile fatty acids production through reuse of acidogenic off-gases during electro-fermentation. *Science of the Total Environment, 668,* 295–302.

Index

CPSIA information can be obtained
at www.ICGtesting.com
Printed in the USA
BVHW011646151122
652008BV00003B/50

9 781032 126173